**How to protect our children from today's drinking culture**

Dr Aric Sigman

For my Little Monkey

PIATKUS

First published in Great Britain in 2011 by Piatkus
This paperback edition published in 2012



A CIP catalogue record for this book is available from the British Library.

ISBN 978-0-7499-5510-6

**Note for the reader**
The recommendations in this book are intended solely as education and information. If you think you or one of your family members has a problem with alcohol, seek medical help from your doctor or equivalent health professional. A list of organisations providing support and further information can be found in the Resources section at the back of this book.

Typeset in Sabon by M Rules
Printed and bound in Great Britain by
Clays Ltd, St Ives plc

Papers used by Piatkus are from well-managed forests and other responsible sources.

MIX
Paper from
responsible sources
FSC® C104740

Piatkus
An imprint of
Little, Brown Book Group
100 Victoria Embankment
London EC4Y 0DY

An Hachette UK Company
www.hachette.co.uk

www.piatkus.co.uk

# Contents

## A note on terms used in the book

While this book is about the *prevention* of alcohol problems, it helps to be familiar with some of the popular terms being used to describe these problems in order to understand some of the studies mentioned.

Alcohol problems are increasingly referred to as alcohol-use disorders (AUDs) which, in turn, are often roughly divided into 'alcohol abuse' (i.e. 'hazardous'/'harmful' drinking) and 'alcohol dependence' (the more politically correct name for good old-fashioned alcoholism – a term still used by many professionals today). In alcohol abuse, a person's drinking leads to problems, but not to a physical addiction. But alcoholism or alcohol dependence applies to a person who shows signs of physical addiction to alcohol (for example, tolerance and withdrawal) and who continues to drink, despite problems with health (physical and mental) and social, family or job responsibilities. In such cases, alcohol may come to dominate the person's life and relationships. While alcohol abuse and alcohol dependence are two different forms of problem drinking, there is certainly overlap between these categories of alcohol-use disorders.

## Acknowledgements

In an age where books on alcohol seem almost exclusively to be penned by celebrities writing about rehab . . . I thank my editor Anne Lawrance and her sober judgment that there *is* another more important aspect to alcohol that has not been addressed. My agent Sara Menguc expedited the creation of this book first by declaring 'This *is* a book', and then by justifiably nagging me to finish my proposal. Moreover, she continues to endure and actively support my unwillingness to write the more comfy, ingratiating 'How to be' books that would probably put many more new pairs of shoes in her wardrobe.

The dispassionate eagle eye of my copy editor Anne Newman was invaluable. She ensured the book still made sense amidst the plethora of studies cited. She even took pity on me and organised my quagmire of citations when I couldn't cope any more.

My colleague, Jonnie Leach, who has for many years lectured pupils on drug education, has been a tremendous inspiration not to mention a great help to me. His raison d'être – 'We're trying to save lives' – is a constant reminder of the underlying significance of what has for too long been thought of as a lifestyle and 'substance-awareness' issue.

I've come to realise and appreciate my parents' influence in cultivating in me a sharpened sensitivity to the way vested interests shape both the research and our understanding of health issues, and that the truth often belongs to those who commission it.

Yet again my wife Katy has endlessly discussed many aspects of the book with me, helping to clarify my ideas. Worse yet, she willingly submitted herself to telling me whether my accounts of ALDH2*2, CYP2E1 and glutamate receptor-signalling genes made any sense to the average drinker.

# Apéritif

The problem of young drinking doesn't take place in a vacuum, but with a backdrop of alcohol-adoring adults including parents, teachers, doctors, police . . . and of course, celebrities.

Even guests in Her Majesty's prisons feel entitled to drink and get drunk. When a violent riot broke out at Ford Prison, an open jail in West Sussex, on 1 January 2011, with fires destroying buildings and staff taking nearly 22 hours to bring the situation under control, it was widely reported that a large number of prisoners were drinking heavily in celebration of the New Year. Apparently the prisoners took great offence at the suggestion that they may possibly be inebriated and should perhaps be breathalysed. But the background to this seems a drawing room – or cellblock – comedy. The local MP and Criminal Justice minister told the BBC that prisoners had been seen drinking in local pubs and hiring taxis to collect 'up to 30 fish suppers at a time'. While a former police officer who served time at Ford Prison a decade earlier, said he knew of one prisoner who 'had a drinks "cabinet" containing bottles of whiskey, rum, Bacardi and gin'. He added, 'when I refused to drink, I was beaten.'

My eyes were acutely opened to modern society's view of alcohol and children when my eldest daughter – fourteen at the time – wanted to come to a health club with me in order to stay fit and healthy, as well as bond with Daddy, I hope. 'It's not allowed Sir. It's against health and safety' came the ironic response from every establishment. On one occasion as we walked away from yet another rejection on a cold dark evening, we passed by a pub in which we could see a boy of about five sitting on a bar stool watching his father, cigarette in mouth while resting his pint on the bar, prepare to play pool. My daughter and I looked at one another. It didn't take long before she asked rhetorically, 'Why is it Daddy, that you can take me into a pub but not a health club?' Good question.

I later discovered that it's perfectly legal for me to serve my children vodka from the age of five.

Culturally, alcohol is in our blood, and for many literally so. But it's increasingly making its way into our children's blood and at alarmingly younger ages. As society wrings its hands over teenage 'binge drinking' we seem unaware that our approach – if we can even call it that – to children and alcohol has been completely at odds with what both medical research and common sense should have told us if they'd been allowed to prevail over the cacophony of mixed messages and drinks-lobby public education.

In order to reduce the likelihood that our children will

develop an alcohol problem now or when they are adults, demands that we first consider young drinking and adult drinking as entirely different matters. As this book will go on to explain, the effects and long-term consequences of drinking even so-called moderate amounts of alcohol are entirely different matters for the young brain, body and genes, with ramifications reverberating for decades beyond a child's first sip.

I've been involved in health education for over two decades. I regularly give talks at schools to groups of children and parents. It's an eye-opener to discover where they get their (mis)information from. This has spurred me to write this book. I had previously felt pretty uninspired about alcohol as a health education topic, assuming it was dreary and predictable. But, in fact, when it comes to alcohol there is a bottomless pit of misunderstanding.

While I realise that in clearing up misunderstanding and folk lore, I will tread on the toes of many adults, this is not only unavoidable, but key to tackling the growing problem our society faces with the young and alcohol. This book is not to be misconstrued as a twenty-first century temperance initiative by another name. It is intended to provide insight and information in order to protect children and young people from harm that alcohol could cause now and in the future. It is intended to garner support for the idea that we establish a widely recognised period of grace for children to develop a greater resilience to the effects of the alcohol so that they may cope better with alcohol as adults. This is nothing more than a straightforward health and development issue.

# One

# It's the Drink Talking

Alcohol is the latest victim in a long line of sensual rationing – preceded by passive smoking, smoking, dietary salt, saturated fat, caffeine, sun tanning, driving $CO_2$-emitting cars and having unprotected sex. In fact, during the 1990s, when concern over AIDS reached mammoth proportions, the government even tried to dissuade the population from engaging in penetrative sex – to give up intercourse and settle for 'outercourse' instead. The Health Education Authority declared: 'Penetration isn't essential. And neither is coming.'[1] Today, alcohol is the latest mainstream pleasure to come under assault; and in a politically correct health-and-safety-obsessed culture, in which drinking is one of the few widely supported conduits for uninhibited behaviour, nobody wants to appear inhibited or uncool.

There is unanimous, if unspoken, political and social consensus over alcohol's role in life. From conservative establishmentarian to student radical, banker and bohemian, it is truly our drug of choice. Musicians and politicians pass out, check in, and dry out … and then take tea together (herbal, of course) in the same celebrity rehab clinics. Celebrity alumni of the Betty Ford Clinic (founded by former US President Ford's wife) include Tony Curtis, Elizabeth Taylor, Stevie Nicks, Johnny Cash, Billy Joel, Drew Barrymore, Liza Minnelli, Kelsey Grammer, Ozzy Osbourne, Keith Urban, Alice Cooper, Jerry Lee Lewis, Tammy Wynette, Robert Mitchum, David Hasselhoff, Richard Pryor and Lindsay Lohan.[2] While *Forbes* magazine has published a guide to the 'Most Luxurious Places To Dry Out'.[3]

And with alcohol being the internationally approved relaxant allowing northern Europeans to let their hair down, society is tetchy about too much criticism of its social leveller, alcohol. Journalists and columnists continue to write witty ripostes to calls for alcohol awareness, brushing them aside as representative of an uptight, interfering, overzealous health-and-safety mentality imposed by anti-hedonistic moralists and killjoys. Pointing out society's alcohol problem is easily dismissed as a puritanical, neo-temperance initiative pushing for neo-prohibition. I suspect this is because these columnists adore drinking, and/or they're 'asked' to write a feature questioning any concern over our drinking. I have no doubt some are actively paid to do so.

For example, emblazoned across the front page of the December 2009 issue of the *Spectator* was the headline, 'Don't worry – drink and be merry'. In the article, the British Medical Association was introduced as 'the puritanical anti-drinking movement' who will simply exacerbate any national problem. Readers were then reassured: 'That's right: drinking is good for us. And I don't just mean half a glass of organic Cabernet with dinner twice a week, either. In fact, there is plenty of evidence to suggest that regular, habitual drinking of the type the government would classify as "heavy" and "hazardous" is significantly beneficial ...'

Further, informed antenatal wisdom is dispensed: 'As social health issues go, it is hard to find a debate more fraught than the one over alcohol and pregnancy. For centuries, pregnant women were encouraged by doctors and midwives to drink occasional, moderate amounts of alcohol in Britain, and no widespread epidemic of foetal alcohol syndrome ensued. Yet what does the government recommend a pregnant woman drinks today? Absolutely nothing ... Strangely enough, this cultural shift flies in the face of recent scientific evidence.' The only two authorities on the matter that the journalist quotes are an American sociologist and a Yorkshire publican and host of a TV show called *Save Our Boozer*.[4]

The article is actually entertaining and thought-provoking, but ultimately the medical claims are the complete and absolute opposite of the truth. The reality is that health professionals expressing concern about the actual documented effects of alcohol have straightforward legitimate grounds for their fears and are not acting on behalf of Jesus. In fact, almost all health professionals drink – some drink a hell of a lot – yet most acknowledge that this is a risk they're willing to take, or at least not think about. What they cannot be involved in, however, is massaging the truth to accommodate modern lifestyles. Especially when children are involved.

## Pub People

Having visited a wide variety of unusual cultures, it's clear to me that drinking occupies a special cultural place in Britain and northern Europe. I only really noticed this when I visited several dry cultures.

Walking around late at night in the centre of Isfahan, Iran, is surreal: people stroll casually; they don't stagger, shout, sway, vomit, urinate or stab one another. In fact, many families walk calmly and happily together. The reason for this, I discovered, is that they're stone-cold sober.

In the Baliem Valley of Irian Jaya (West Papua) I found myself with a tribe of essentially 'stone-age' people. The men had spears (big ones) and were naked except for penis gourds (very big ones) and massive boar tusks through their noses. The chief introduced me to his four wives, told me that the chief in the next valley had ninety-six and asked how many I have. 'Just the one,' I replied. Pointing to the compound entrance, the chief informed me that they would only spear someone if they came into the village to steal their pigs (their currency) or their wives. (The going rate for a decent wife was, I was told, ten to fifteen pigs. And I witnessed a mediation session in which the lover of an adulteress was fined sixty pigs to be paid to the cuckolded husband.) There was talk of past cannibalism of tribal enemies. Yet you couldn't – and I have never – meet a nicer group of people. They were both attentive and considerate and you'd be far safer among them than on your local high street on a Friday or Saturday night. And they're sober too. But we apparently refer to them as an 'uncivilised'.

Back here, in many communities, the pub is treated as the neighbourhood sitting room, where children are welcome. Even underage teens can go into a pub by themselves, sit and drink soft drinks. In addition to this, there is a general belief that the early 'responsible' introduction to drinking at home will in some way prevent heavy drinking later. And because drinking is such an integral part of adult life, we are uneasy when this comfortable, 'sensible' view of drinking is turned upside down and challenged. But a new crop of studies is doing exactly this, and you may need a stiff drink before you continue reading.

Adolescents and teenagers have traditionally experimented with alcohol, but what we've seen since 2005 far exceeds mere rites of passage. Even hardened journalists – traditionally a profession not known for an aversion to drunkenness – are increasingly aghast at the sheer degree of drinking in our society.

On returning to Britain after ten years working as the BBC's North America correspondent, Justin Webb was moved to write, 'But nothing prepared me for the booze. Sometimes it seems as if everyone is drunk ... the most striking feature of the return across the pond: America is sober, Britain is legless.'

In describing an upper-echelon Washington media party where alcohol wasn't served, Webb compares and contrasts what constitutes a good time: 'Getting tipsy with friends – one of the great British pastimes ... In sober, temperance America this [cherryade] is what passes for a good time.' And even among journalists at work the clash of liquid cultures is unavoidable. Webb describes a friend who worked for Sky News in Washington. In order to get to his bureau he had to walk through the Fox News office, so to be friendly on his first day, he called out that anyone who fancied a drink should come with him. It was lunchtime. 'He might as well have told the (generally right-wing) Fox folk that he was a socialist transvestite. He was worried, he told me later, that they were going to call the cops.'[5]

Another even more hardened journalist, John Humphrys, returning to his home city to observe a typical Saturday night under the protection of police in command of Operation Cardiff After Dark, confessed, 'What I experienced the last time I was there a few days ago shocked me – and it might have been almost any city centre or small town across the land on any Friday or Saturday night … Everyone I spoke to told me the same thing: "We're here to get drunk."' Every police officer he interviewed from commissioner to bobby on the beat felt strongly that, 'There are simply too many pubs, and far too many of them are so-called "vertical drinking establishments". There is nowhere to sit and chat over a quiet pint, nowhere even to rest your glass. You stand and drink. The whole point of the evening is to get drunk.'[6]

## The State We're In

Statistics have their limitations and, for many of us, they make dull reading; but they also have their uses in bringing some clarity to what has often been a cultural or moral issue. Most, if not all, other countries have had increasing problems with underage drinking, but Britain is a particularly good case in point.

A report from the Organisation for Economic Co-operation and Development found that young teenagers in the UK are more likely to get drunk than their counterparts anywhere else in the industrial world. The UK's figure for underage drunkenness – 33 per cent (measured in terms of the proportion of thirteen- to fifteen-year-olds who have been drunk at least twice) – is more than double the rate for countries such as the United States, France and Italy. Among girls, the gap between the UK and other countries is even wider.[7]

While the National Health Service reported a slight increase in 2009 in children who had never drunk alcohol, those who do drink alcohol seem to be drinking even more.[8] In England, eleven- to fifteen-year-old children drink an average of 14.5 units of alcohol a week; in the north-east almost 18 units – equivalent to nine pints of beer or one and a half bottles of wine.[9]

And there are more frightening statistics:

- The number of children being treated in hospital A&E departments because they have drunk too much has risen sharply, according to the NHS Information Service. Hospital figures show a 32 per cent increase in four years, with more girls needing treatment than boys.[10]
- The number of teenagers hospitalised for alcohol poisoning has risen dramatically since 2003; and, for the first time in history, girl patients outnumber boys by 3:1.[11]

- The increase in drinking among twelve- to fourteen-year-olds has led to the highest rise in rehab admissions ever reported – up by 62 per cent in a single year.[12]
- The notion of 'spiked drinks' and 'date-rape drugs' is now being described by scientists as 'an urban myth': in almost all cases, victims are actually found to have only high amounts of alcohol in their blood – nothing more.[13]
- One-third of teenagers killed on their bicycles have been drinking alcohol.[14]
- Three-quarters of pedestrians, including adolescents, killed by cars have been drinking.[15]
- Up to seven out of ten killings, stabbings and beatings are directly related to alcohol.
- As a result of teen drinking, the Royal College of Physicians reports it is now commonplace to see young people dying of liver disease in their twenties. NHS figures (2011) show that the number of young drinkers with serious liver problems has risen by more than 50 per cent in the last decade.

In late 2010, a survey commissioned by Drinkaware and supported by the British Home Secretary found that more than a third of eighteen- to twenty-year-olds go out with the explicit intention to get drunk. More than one in four reported they had no idea how they got home at least once in the last year; one in three thought it was acceptable to wake up not knowing how they got home after a drinking session; and one in twenty-five believed it was acceptable to end up in hospital.[16]

## Drink to Your Health?

NHS figures from late 2010 show that one person every seven minutes is admitted to hospital in England for a health problem directly attributed to alcohol. The number of alcohol-related admissions increased by 54 per cent in ten years. This excludes conditions that are merely exacerbated by drink, such as diabetes and most types of heart disease. It includes people who have had too much to drink and are classified as suffering anything from alcohol poisoning to cirrhosis of the liver, as well as some admissions from A&E and some directly into hospital via ambulance or doctor's referral.[17]

Generally, people are now admitted to hospital for alcohol-related diseases at a rate of almost 1 million per year. However, when figures for patients injured while drunk or victims of alcohol-related violence are added, this number is far higher.

The House of Commons Health Committee has recently summed up alcohol's role in the state we're in: 'The fact that alcohol has been enjoyed by humans since the dawn of civilisation has tended to obscure the fact that it is also a toxic, dependence-inducing teratogenic [causing malformation] and carcinogenic drug to which more than 3 million people in the UK are addicted. The ill effects of alcohol misuse affect the young ... Alcohol has a massive impact on the families and children of heavy drinkers, and on innocent bystanders caught up in the damage inflicted by binge drinking. Nearly half of all violent offences are alcohol related and more than 1.3 million children suffer alcohol-related abuse or neglect. The costs to the NHS are huge, but the costs to society as a whole are even higher, all of these harms are increasing and all are directly related to the overall levels of alcohol consumption within society.'[18]

One can quibble about statistics showing a slight upward trend this year or a marginal decline that year, but only a fool would refute that we have a substantial problem.

## Parental Role Models

Beyond the hard statistics lies the emotional landscape of how children feel when they see their parents drinking. A report in 2010 by ChildWise into the impact of adult drinking on children recently found that nearly a third feel scared when they see adults drunk or drinking too much. Half of the 1234 ten- to fourteen-year-olds questioned said they had seen their parents drunk. About one child in every class of about thirty said they saw their parents drunk several times a week. Eight out of ten children who had seen adults drinking said they noticed a change in the way they behaved; of those, almost a quarter said alcohol made adults 'act stupid or silly', while a fifth said they became 'angry and aggressive'.

Nearly half of the children questioned were 'not bothered' by drunkenness, which, according to the report's research director, suggested drinking culture had become 'ingrained'. Girls aged over eleven were asked why adults drank until they lost control and vomited. Several responded by saying that it was 'part of a good night out'.

One of the report's authors commented, 'It's going to make it very, very difficult to tackle the culture of binge drinking; it's so embedded in their idea of what's a good night out.' The report indicated that children need to be educated about alcohol from an early age if binge drinking was to be tackled effectively and warned that action would be needed to prevent it from worsening among the next generation: 'Each generation takes the cues from their parents, and if their parents are drinking more it makes it easier for them to drink more.'[19]

The British Liver Trust too is concerned, saying: 'The influence that parents have on their children's drinking is incredible and something that shouldn't be overlooked. Sadly, we know that increasing numbers of young people are suffering serious health problems, including fatal liver damage, due to drinking too much alcohol.'[20]

A major government study published in 2010 found that young drinking is a particularly 'middle-class problem' and that affluent, liberal families are most likely to follow the 'continental' practice of letting children have a small amount of drink.[21] 'In general, white parents, heavier drinkers and AB parents [the wealthiest social group] were more likely to think that it is acceptable for children to start drinking at a younger age.' While a related government study adds, 'This seems to indicate that young people of a very low social position may be less likely to try alcohol, possibly because it is less available in the home.' Researchers found 62 per cent of children regularly drank alcohol at home and a quarter said they had been drunk. Parents underestimated the amount they themselves drank, saying they were 'light' or 'occasional' drinkers, when they actually consumed a great deal more. Parents also underestimated the impact of their own drinking habits on their children's attitude to alcohol, while other studies show that this has a powerful influence. A quarter of young people have never spoken to their parents about drinking alcohol. In January 2010, the Department for Children, Schools and Families (called the Department for Education since May 2010) urged parents, children and young people 'to have open conversations about alcohol, to ultimately delay the age at which young people start drinking'.

But many parents are not exactly sure what to say. Most adults are understandably confused because they are bombarded with sensational headlines, misinformation and conflicting messages about the risks and benefits of alcohol, and the best way to prevent our children from drinking too much. For example, while some studies suggest that small amounts of alcohol can reduce the risk of coronary heart disease, few of us realise that these findings only apply to *men* over the age of forty, if at all. And that for women, the level of alcohol consumption with the lowest risk to death is zero.[22] While we admire the French, whose 'continental approach' to drinking we assume is healthier than our own, France's death rate from cirrhosis of the liver has until recently been *twice* that of England and Wales.[23] While we increasingly use the term 'binge drinking' to describe the visibly drunk, how many of us realise that the Royal College of Psychiatrists clearly defines 'binge drinking' as the consumption of three small (125ml) glasses of wine for a fully grown adult woman and four for a man 'in a day'.[24]

Drinking is something we enjoy and inevitably see as an integral part of adult life. Unless we or someone we know has an alcohol problem or we have to drive home after a dinner party, alcohol is something we rarely think much about. Books on the subject tend to fall into two categories: coping with alcoholism and good wine guides. We think more carefully about our children and GM food, passive smoking, health and safety, Internet predators, paedophilia and, of course, drugs. But to confront underage drinking requires us to reflect upon our own attitude to alcohol. The way we view alcohol has a direct impact upon our children's present and future health, wellbeing and academic success.

And while binge drinking is seen as a particular social problem that increasingly makes the news, beyond the headlines is a growing body of empirical evidence about to force us to reconsider our relationship with alcohol, for our own sake and our children's too.

## Early and Often to Avoid Disappointment

Despite the direct links between alcohol and big problems for our children and their behaviour and attitude towards authority, it is only recently that the penny has begun to drop. Even doctors and nurses that I know have been carried along with the assumption that the best way to prevent our children from drinking heavily and behaving badly as a result is to teach them to drink while they are young. This belief has been heavily promoted by educational bodies that appear to be impartial, but are funded by the drinks industry, well-versed in the comfy speak of 'teaching children sensible drinking' or 'responsible drinking'.

In 2009, an English prep' school was reported to have invited a local wine merchant to give forty twelve-year-old pupils a class in wine tasting entitled, 'Wines of the World', during which children tasted various wines. Isn't it interesting that even if it was legal we wouldn't recommend early sensible snorting to prevent later cocaine addiction and abuse – or sensible spliff/dope smoking to prevent later cannabis abuse, early cigarette smoking to prevent later nicotine addiction or early sexual encounters to prevent teenage pregnancy? Yet when it comes to our logic regarding introducing children to alcohol, we seem to be thinking under the influence of alcohol. Our complacency sees child and teenage drinking as inevitable, yet this is only inevitable because we not only allow it to be, but we inadvertently make it so.

It's time we had a cultural look in the mirror.

## Swallow, but Don't Inhale

For a nation with a perpetual hangover, Britain's attitude towards drugs has been reasonably priggish. Prime ministers and wannabes are concerned about whether the media will discover or even ask, 'Did you inhale a quarter of a century ago when you were at Oxford?' At the same time, most prime ministers and cabinet ministers would be embarrassed if the media accused them of being stone-cold sober during their entire university years and their spin doctors would be called in to create retrospective socially acceptable anecdotes of jolly user-friendly drunkenness experienced by the in-touch-with-today's-youth politician.

Adults love alcohol and governments collect extraordinary levels of sin tax from it, all the while pointing to drugs as the greatest menace to our children and their behaviour. Yet alcohol is, always has been and continues to be, by far, our children's greatest drug problem. At the end of 2010, a study published in the *Lancet* ranking twenty of Britain's most popular drugs according to their degree of harm placed alcohol at number one: 'Drugs were scored out of 100 points ... Overall, alcohol was the most harmful drug (overall harm score 72), with heroin (55) and crack cocaine (54) in second and third places.' Alcohol is now considered far more harmful than *all* popular illegal substances including ecstasy, LSD and cannabis. And here are some of their scores: crystal meth (33), cocaine (27), amphetamine/speed (23), cannabis (20), GHB (18), ecstasy (9), LSD (7) and magic mushrooms (5).[25]

And while everyone still remembers the name Leah Betts fifteen years after her death, and the campaigns telling children that 'ecstasy kills', there is no equivalent for alcohol poisoning – after all, can you recall hearing the name of any adolescent in the news, even though there are, in fact, tens of thousands of them? And there are no campaigns in any one of *their* names. (Subsequently, it was discovered that the direct cause of Leah Betts's death was actually 'water intoxication', not drugs – but that's not what society wanted to hear or believe.) The Royal College of Psychiatrists summarises the hypocrisy nicely in their publication 'Alcohol: Our Favourite Drug', where they tell us:

'Alcohol causes much more harm than illegal drugs like heroin and cannabis.'[26]

One of the most telling signs that alcohol really is an enormous issue in child development is the observation that states of the United States are willing to forgo the tremendous amount of sin tax they would get from alcohol sales by raising the legal drinking age from eighteen to twenty-one. That should tell us all something.

## Changing Our Culture

We need to take a cold, hard, detached look at the cultural background that has prevented us from protecting our children from alcohol and the way it affects their social development, behaviour and attitude to authority. This re-examination is not going to be easy. Part of the reason is because we're too close to the subject matter; so close, in fact, that many of us will have a significant amount of alcohol in us by tonight. And the same goes for our children's heroes. I routinely hear radio DJs joking about how hammered they were the night before and taking emails, texts and phone calls from freshly hung-over youths.

The non-drinker is now considered so uncool and anally retentive that not drinking is akin to spending Friday night in the library or in church, God forbid. There's something wrong with you if you don't want to drink or if you don't want to go somewhere where most people get drunk. And the interesting thing about this kind of conformity is that it's supported by the most eclectic group of bedfellows, from left-wing university-subsidised bars to the House of Commons, where our MPs have a choice of twelve bars subsidised by us taxpayers and where, apparently, when Peter Mandelson resigned for the second time, 'the bar ran out of champagne'.[27]

You couldn't make this up. And I haven't. Indeed, as I was writing this chapter, my wife handed me a news story just breaking, 'MPs drunk as they voted on the budget'. Apparently, dozens of MPs were drunk, with one cabinet minister reported to be slurring his speech:

> 'It was disgusting,' said a female MP. 'The Chamber and the voting lobbies stank of booze and sweat.' Another added: 'Several people were legless. MPs old enough to know better were all over the Sloane Rangers who have come to work here as secretaries and researchers since David Cameron got in.'

One MP was too drunk to vote after he fell to the floor of a House of Commons (subsidised) bar and made a public apology on the BBC for his irresponsible actions. His name was the Rt Hon. Mark Reckless.[28]

Journalists were quick to investigate drinking establishments, brandishing their calculators to compare the House of Commons with high-street pub prices. Apparently, while a small glass of Pinot noir costs £3.85 in a typical pub in Birmingham, it's a snip at only £1.80 at a House of Commons bar. The House of Commons Refreshment Department tell me that 'bars in the House of Commons operate without a licence, and do not keep to the permitted hours laid down by the Licensing Acts ... because Members and staff of the House require refreshment of all kinds whenever the House or its committees are sitting ...'. The bars come with a variety of folksy names, such as Annie's Bar and the Strangers' Bar, where – unfortunately for the Rt Hon. MP above – everybody knows your name.[29]

From what I have seen, Labour MPs don't want to disapprove of drinking because they market themselves as champions of the real people – and 'real' people spend a lot of time in pubs and clubs, drinking being the national pastime and traditional escape from the shackles of the capitalist class system. And the Conservatives see any preoccupation with the negative effects of drinking as lower middle-class pedantry and, as such, decidedly uncultured.

The problem with conducting a sober discussion about alcohol's role in our children's socialisation has been that Britain doesn't really have its heart behind this issue, such is its people's long-standing affection for the nation's silly sauce.

Fortunately, however, there may be a slight change in the air. In 2009, the British government's Health Secretary at the time acknowledged: 'Non-drinkers are often subjected to the same disdain that non-smokers were thirty or forty years ago when people looked at you strangely if you refused a cigarette. They are the odd ones out ... the question we must ask as a society is why, unlike smoking, it is the abstainers that draw people's attention, not those who regularly drink their weekly limit in a day ... If we want to make further progress, these cultural questions must be addressed.'[30]

## A Class Act

One of the advantages of being a foreigner in Britain is that I am, to some extent, socially chameleonic. I have no doubt that if I had the middle-class, non-estuary accent of my English counterpart I wouldn't be privy to many of the things I've observed and experienced in Britain's class jungle. One such observation is how, as Britain has supposedly become a more 'classless' society, the authorities have enabled the 'lower orders' to subjugate themselves. Every government realises that too much thinking on the part of the masses can at times cause discontent and unrest, and even a loss of an election. So by liberalising drinking laws, encouraging an explosion in the numbers of entertainment television channels and now doing deals with broadband providers to ensure every man, woman and child has unlimited broadband access (even though it is well documented that the Internet is rarely used for educational or political purposes) the class system is perpetuated. Everyone's either tipsy in the evenings or watching a different lifestyle channel or website – or both at the same time. It's not surprising that there's far less political activity among the young – or even the old, for that matter: distract + anaesthetise = divide and rule. The cleverest part of this is that the public see these things as new privileges. I also recall the public waiting at budget time to see if the chancellor would be putting another penny on a pint. And it can only be a coincidence that in 2010, when the new government unveiled the most severe spending cuts since the Second World War, alcohol was left untouched by the tax man.

But while the liquid bread and circuses may spare a government too much independent analytical thought and possible political action on the part of the electorate – it does, unfortunately, enable low-status man (and increasingly woman – 25 per cent of arrests for violent assaults are now of females) to get in touch with his inner class hatred, leading to random violence, murder, wife beating and other forms of self-expression impervious to anger management. It has been a trade-off for the powers that be, who constantly try to juggle the cost/benefits of public pacification through the Devil's buttermilk.

This may seem like a political tangent, but it's important to understand the cultural background that has prevented parents and policymakers from protecting children from alcohol and its effects.

## Binge Britannia

Of course binge drinking in Britain is nothing new. Let's take a look at some of the nation's liquid highlights that the history syllabus does not dwell upon.

### Edwardian inebriation

In 2010, the Birmingham Pub Blacklist was launched online, detailing the drunkards whose loutish behaviour saw them barred from the city's pubs and clubs at the turn of the last century. Serial drunks were placed on the list after receiving four convictions under the Inebriates Act of 1898, which included being intoxicated to the point of complete incompetence (a bit like some of our modern-day parliamentarians) and being found in a shebeen – a bar that sold alcohol without a licence. Other offences included riding a horse while under the influence or drink-driving a steam engine.

The information was compiled by the Watch Committee of the City of Birmingham, set up by the police to enforce the Licensing Act of 1902 which was passed in an attempt to deal with public drunks, giving police the power to apprehend those found drunk in any public place and unable to take care of themselves. The Blacklist provided licensed liquor sellers with photos and descriptions of 'habitual drunkards' who were not to be sold alcohol due to their reputation and past delinquencies.

Each drunkard's entry includes photographs (front and profile views), their name, alias, residence, employment (including prostitute, 'bedstead polisher', 'hawker' and 'grease merchant'), a physical description, any distinguishing marks (such as tattoos and scars), the nature of their conviction and the sentence received for booze-related crimes.[31]

## A stroll down Gin Lane

The production and consumption of English gin, otherwise known as Mother's Ruin, was then popular among politicians long before they developed today's penchant for taxpayers'

Chardonnay. Even Queen Anne was actively encouraged by the government to drink up. As Charles Davenant, an English economist at the time, noted: 'Tis a growing fad among the common people and may in time prevail as much as opium with the Turks.'

And so the Gin Craze in the first half of the eighteenth century was helped by its new-found popularity among the working classes. In the slums of Clerkenwell, gin was in such demand that it was sold from wheelbarrows in the street. By 1721, however, Middlesex magistrates were already decrying it as 'the principal cause of all the vice and debauchery committed among the inferior sort of people'. In 1729, Parliament passed a Gin Act that increased the tax on the drink. Our current government may wish to recall that this was unpopular with the working classes and, in 1743, resulted in riots in London. Coincidentally, one loophole used by retailers to avoid tax was the sale of the spirit under the pseudonym 'Parliament gin'.

England's inebriety was ingloriously highlighted by Hogarth's engraving 'Gin Lane', featuring – quite literally – the consummate slummy mummy of the day sporting an expression of blissful ignorance, while seriously neglecting her child as it falls to its death. The idea was more recently updated as 'Cocaine Lane', with drugs being substituted for gin and a scene featuring loft conversions, wine bars and mobile phones.

'Gin Lane' was created in conjunction with another piece entitled, 'Beer Street' in which its up-town inhabitants are portrayed as happy, healthy and nourished by the lovely English ale – beer being promoted by many anti-gin campaigners as the patriotic (and sober) alternative to gin. Despite this, beer consumption fell significantly throughout the eighteenth century, largely due to the increasing popularity of the new 'soft drugs', including coffee, tea and chocolate.

## How the West was *really* won

The United States of America came into being because of a rebellion – the Boston Tea Party – over the British government's tax on tea. And the first major challenge to the authority of the new US government came in the 1790s in the form of the Whiskey Rebellion. Pennsylvania rebels used violence and intimidation to stop federal officials from collecting alcohol tax, as more than five hundred armed men attacked the fortified home of tax inspector General John Neville. President George Washington himself led the army to suppress the rebellion, sending an important message establishing a fundamental principle: that the new national government was both willing and able to suppress violent resistance to its laws.

Interestingly, alcohol played a significant role in relieving the American Indians of their furs and their lands. Whiskey was used by the Michigan Department of Indian Affairs[32] when the US Commissioner for Indian Affairs, Thomas L. McKenney, observed that in their original state the Indians of the Great Lakes were noble savages, but that 'the curse has been inflicted that has ever since been wearing down the population of the once mighty population … spiritous liquors'.[33] Not all tribes were susceptible to the liquor trade, but for those that were, it created widespread havoc. The Indians willingly accepted traders' whiskey and soon reached the point where they would not trade without it. Once this inherent weakness for alcohol was made apparent, fur traders and land speculators used it to get their furs and land.[34]

The discovery of distant lands that Britain would eventually colonise was virtually dependent on drunkenness. Captain James Cook lived in fear of uprisings on board if liquor provisions ran low, meaning that sailors had to be asked to accept a temporarily reduced allowance. In fact, Cook actually planned and co-ordinated his expeditions according to the state of the liquor reserves. The daily ration of grog was one imperial pint (20 fluid ounces or over half a litre of 94 per cent proof liquor). Cook would not cruise for longer than the liquor lasted.

Then there was the traditional ceremony of crossing the equator, whereby each new initiate could either pay a fine in alcohol (to sacrifice four days' ration of liquor) or be tied up and then ducked three times in the ocean. Lieutenant John Gore wrote, 'Many of the men, however, chose to be ducked, rather than give up four days' allowance.' And some, he reported, 'almost suffocated'.[35]

Cook's lieutenant John Williamson wrote, 'A seaman would as soon part with his life, as his grog.' And when, on the east coast of New Zealand's North Island, Midshipman John Rowe was confronted by a large hostile crowd of Maori, he wanted to capture and hold a few of the 'savages' hostage to ensure the return of a stolen cask of liquor, despite the extraordinary danger this presented.

## G&T timeline

Striking images of drunkenness such as Hogarth's 'Gin Lane' have encouraged the widespread belief that the English have always been a nation of drunks. Far from being a story of perpetual drunkenness, however, English drinking habits have fluctuated wildly over the years, beginning with a long decline in alcohol consumption from the late seventeenth century. There was a blip in the first half of the eighteenth century associated with the gin craze, and then increasing levels in the mid-1800s, but these fell rapidly and significantly later in the century. Levels in the inter-war years were low and remained so until the 1960s.

But the last half-century has seen a significant shift in these trends, not just because consumption levels have risen again, but also because of the growing popularity of stronger drinks – in particular wine and spirits and, more recently, strong white cider. We now drink about three times more per head than in the years of lowest consumption – a trend generated by the same factors which influenced drinking in the past, namely, economics (affordability and availability) and changes in culture.[36]

In some ways, little has altered since an anonymous Chinese commentator writing around 650BC asserted that people 'will not do without beer. To prohibit it and secure total abstinence from it is beyond the power even of sages. Hence, therefore, we have warnings on the abuse of it.' What is different today is that our warnings are more specific and qualified than ever before.

In their 138-page report from 2010, entitled, 'Alcohol', the House of Commons Health Committee summoned a large number of expert witnesses and reviewed many current studies in order to take an historical look at alcohol – a patient history – before prescribing future treatment:

> One of the biggest changes in the last 50 years has been in the drinking habits of women and young people:
>
> > Whatever their social and cultural standing – i.e. Ugandan 'youths', medieval knights, the Victorian urban 'poor', 20th-century 'post-modernists', 16th-century 'wits', Somali village elders – drinking, especially to excess, has been a masculine preserve ... What is striking about current trends in Britain is that women are now engaging in many of the same drinking practices as men, and consuming similar if not more amounts of alcohol in the process.
>
> Teenagers drink twice as much as they did in 1990.[37]

## In Denial and a Conflict of Interests

In trying to understand how we have allowed things to continue in this way there are plenty of interesting co-conspirators to consider. The National Union of Students has argued that 'students' unions are some of the most responsible retailers of alcohol'. Yet it then admits that unions needed to keep up alcohol sales in order to fund student services, and that this leads to drinks promotions and, consequently, binge drinking and antisocial behaviour.[38]

The heads of Universities UK (the representative organisation for the UK's universities) insist that when it comes to alcohol, universities 'did not have a duty of care for their students'. However, the House of Commons Health Committee says that while the National Union of Students and the universities themselves appear to recognise the existence of a student binge-drinking culture, 'all too often their approach appears much too passive and tolerant ... The first step must be for universities to acknowledge that they do indeed have a most important moral "duty of care" to their students, and for them to take this duty far more seriously than they do at present.'[39]

Even Varsity Leisure Group Limited, owner of the Carnage UK brand which has, unfairly or otherwise, become a notorious example of a promoter of nightclub events for students, has given evidence that 'students are being immersed into a culture which is focused around the culture of alcohol. The culture may need to change; the offering of cheap drinks promotions and alcohol-led events may need to be addressed.'[40]

The Royal College of Physicians has now reported that, 'In the UK, the health harms caused by alcohol misuse are underestimated and continue to spiral: ... 6.4 million people consume alcohol at moderate to heavy levels (between 14 and 35 units per week for women and 21 and 50 units per week for men) ... In the last thirty years of the twentieth century, deaths from liver cirrhosis steadily increased; in people aged 35 to 44 years the death rate went up eight-fold in men and almost seven-fold in women.'[41]

The House of Commons Health Committee has come up

with some pretty unambiguous punchlines, a few of which I've highlighted in bold: 'We believe that England has a drink problem. Three times as much alcohol per head is drunk as in the mid-twentieth century. It is **not** just a problem for a small minority, for the obvious alcoholics and heavy binge drinkers, but for **a much larger section** of the population. Ten million people drink more than the recommended limits, 2.6 million more than twice the limit ... Twenty-seven per cent of young male and 15 per cent of young female deaths were caused by alcohol. **Our teenagers have an appalling drink problem** ... The alcohol problem in this country reflects a failure of will and competence on the part of government Departments and quangos.'[42]

We are then given a rare glimpse of how important government reports are massaged by politicians to prevent us from becoming overexcited about alcohol and our children.

In March 2004 the government published the 'Alcohol Harm Reduction Strategy for England'. In the Foreword, then Prime Minister Tony Blair wrote: 'Increasingly, alcohol misuse by a small minority is causing two major and largely distinct problems ...'. However, the House of Commons Health Committee was astounded by Blair's words which said the exact opposite of the medical facts in the report for which he was writing the Foreword: 'These comments were surprising ... since the report showed that misuse was **not** a problem for a **small minority**: it stated that "a quarter of the population drink above the weekly guidelines of 14 units for women and 21 units for men. It also observed that 5.9 million adults were "binge drinking".'[43]

A clear case of sexing down the documents.

In 2009, the Royal College of Psychiatrists told the House of Commons that a major study by the World Health Organization (WHO) showed the most and least effective alcohol policies across the world. But when comparing these findings with its own Alcohol Harm Reduction Strategy for England, it was clear that the government 'eschewed the most effective policies and adopted the least effective'.[44]

The Commons Committee does the decent thing by laying bare the pretence of 'social responsibility', which the drinks industry increasingly tries to project. 'If everyone drank responsibly the alcohol industry would lose 40 per cent of its sales and some estimates are higher. In formulating its alcohol strategy, the Government must be more sceptical about the industry's claims that it is in favour of responsible drinking.'

The House of Commons Health Committee believes, 'Moreover, people have a right to know the risks they are running.' However, alcohol-awareness campaigns are poorly funded and ineffective at conveying key messages. Not so with the breweries' messages though: as long ago as 2003, the alcohol industry was spending between £600 and £800 million on marketing. It really is the drink talking.

And it's time we learned to talk back.

## Two

# They Tried to Make Me Go to Rehab

In trying to understand why children and young people are drinking more and earlier, it becomes clear that there isn't a singular reason. It's more a case of a variety of disparate causes coinciding. Celebrity culture, electronic media, sexual politics and changing family structure and dynamics all conspire, under the heading 'modern life', to coax our children towards the bottle.

Simple factors are the cheap price of alcohol and the amount of 'discretionary income' (formerly known as pocket money) children have at their disposal.[1] But one of the most important reasons, particularly for girls, has been the inversion of the taboo/stigma previously attached to being smashed out of your mind. Drinking, especially to excess, has across the world been traditionally a masculine preserve. But girl power has had some unintended consequences for a new generation. The Home

Office reports that in only two years, the number of sixteen- to seventeen-year-old girls fined for being drunk and disorderly has risen by nearly 50 per cent, while the rates for their older sisters (eighteen- to twenty-year-olds) have only risen by a mere 40 per cent. The rate of fines is now increasing faster than that for men. And the sexual equality influence gets better and better: women are now involved in a quarter of all violent attacks, and a high proportion are drunk.[2]

A short while ago, celebrities would have paid handsomely to prevent news of their inebriation or admission to rehab from reaching the public's ears. All that's changed now. No satirist could have created a spoof that captures more brilliantly the irony of the story of Britain's greatest soul singer – a nice Jewish white girl from the comfy north London suburb of Southgate: the girl develops an alcohol problem, her management team encourages her to go to a rehabilitation clinic and she refuses. Instead, she pens a song and records it. The song wins the highly prestigious Ivor Novello Award for Best Contemporary Song and three very hard-to-come-by American Grammy Awards, for Record of the Year, Song of the Year and Best Female Pop Vocal Performance. The song is entitled, 'Rehab', the heroine is called Amy Winehouse and she's attracting great interest from some of the most unlikely sources. In addressing alcohol misuse among young females, the president of the Royal College of Physicians, Professor Ian Gilmore, said: 'Even *I've* heard of Amy Winehouse ... and the "Amy Winehouse factor" isn't helping the situation.' The professor will be pleased to hear, however, that after initially saying, 'No, no, no' to the suggestion of rehab, and, 'I won't go, go, go', Ms Winehouse is singing a different tune which goes something like, 'Yes please, what's the address?'

Female models and pop stars are increasingly photographed legless, fighting and even vomiting outside nightclubs, while advertisers enculturate children and young people with notions of alcohol coolness. And so health professionals are now formally recognising the effect of celebrities, otherwise known as 'role models', in having created an alcohol chic. There are even celeb rehab websites listing who's in and who's out in the rehab charts in any given week. Parents may be surprised to hear that some of their celebrities have spent years in and out of rehab. In her autobiography, *Little Girl Lost*, Drew Barrymore talks about a descent into alcohol problems that began when she was nine; her mother put her into rehab and she emerged sober at fourteen.

## Alcohol Chic

Between them, celebrity culture, sporting heroes and role models have created an alcohol chic.

In 2008, research, funded by the UK's Economic and Social Research Council, on eighteen- to twenty-five-year-olds confirmed the obvious – that celebrities renowned for their party lifestyles (including models, such as Kate Moss and some members of the girl group Girls Aloud) were repeatedly cited as evidence that drinking to excess could be attractive and posed few risks. Electronic media, including social networking sites, were seen to reinforce the connection between fun and heavy drinking, the researchers said, because many young people used them to exchange images of alcohol-fuelled exploits. The research involved an analysis of more than two hundred advertisements for alcohol. One of the authors commented that images of celebrities spilling out of nightclubs reinforced the idea that drinking was 'cool' and harmless:

> What was clear from our research was that a good night out involved 'determined drunkenness', which meant planning to get drunk as part of a regular big night out and 'pre-loading', by drinking heavily before leaving the home ... They look at magazines like *Heat* showing celebrities after too much to drink; they talked a lot about people like Kate Moss, and they see extreme drinking as part of a glamorous lifestyle. These are the messages that are getting through; the ones about how many units you can safely drink are not even in there.

The researchers said the appearance of celebrities photographed drunkenly emerging from parties and nightclubs have convinced many young people that binge drinking carries few dangers. The young people said things like: 'They are still looking good and fit; it shows you can drink and get away with it.' They don't think of life in terms of balance, as older people might do, but more in terms of extremes.[3]

Beyond the age, the amount and frequency, there is nothing new about boys and drinking. Strength, performance and manliness have been implied, if not guaranteed, in media messages about alcohol. Curiously, erectile dysfunction doesn't feature in any Carlsberg adverts. And even more curious still, few teenage boys are aware of Harry Redknapp's virtual ban on Tottenham Hotspurs players drinking at all during the football season, saying, 'Footballers should dedicate their lives to playing.' As Redknapp says, 'You shouldn't put diesel in a Ferrari. I know it's hard, but they are earning big money; they are role models to kids.' Boys look up to different sportsmen and the impression they leave is deep and long-lasting. But beyond the responsibility issues lies performance. Alcohol impairs fitness, reaction time and the players' ability to be at the top of their game. In a performance-related age of showbiz TV football, if a player is earning £70,000 a week, it's hard to justify to shareholders the young bucks pumping substances into their bodies that affect their on-field results. In America, where a lot more money is at stake, the National Football League (NFL) has banned clubs from serving alcohol at team functions or on buses or flights. The ban applies not only to players, but also to owners, coaches and guests. Interestingly, few boys are aware that David Beckham and Muhammad Ali are consistently cited as being non-drinkers.

Analyses of TV, radio, magazines and the Internet find that many adverts associate drinking with 'having a laugh' and as integral to the ability to socialise. Media images of alcohol, whether outright adverts, drama or reality documentaries, increasingly highlight all-female friendship groups, carrying a 'Girls just wanna have fun' message. One researcher describes 'the promise of sociability as a treasured prize that is inextricably associated with a "good night out". The ultimate prize on offer is to belong.'[4] Female friends are seen to let their hair down and behave outrageously – safely among themselves. Boys have always been portrayed as bonding with 'mates', while going out into the world to exhibit TV-friendly managed laddishness and have exciting adventures. Drinking is seen as enabling and enhancing your ability to connect.

In the 1950s and 1960s a cigarette casually tucked between the lips of the young rebel (epitomised by actors like Marlon Brando and James Dean or the Marlboro Man) led many boys to light up. But in the late 1990s, social engineering and clean-living Tom Cruise finally relegated smoking to the level of the unhealthy lower-rent loser, reeking of cigarette smoke and lack of motivation and drive, especially in anti-smoking campaign ads, ultimately arriving at the phallocentric warning of today: 'The more you smoke, the less you poke'. While celebrities have always enhanced the appeal of smoking and drinking, it's the intensity of this effect that has changed. The sheer number of opportunities for younger and younger children to see influential role models doing everything from sleeping to blowing their noses in public has burgeoned. The number of celebrities visually available in a child's life is incomparable to that of a decade or so ago. In addition to a teen edition of *Now* magazine, multiple screens (websites, television, DVDs, YouTube, videophones) deliver the highest dose ever of 'role models' directly to the bedrooms of children. And when it comes to the effects of electronic media, the dose matters. In most studies assessing the effects of screen time on children's wellbeing there is a dose–response relationship: the more hours they spend, the stronger the effects.[5] And overdosing on role models bears some relationship to overdosing on alcohol.

A major international meta-analysis of thirteen long-term studies following up a total of over 38,000 adolescents and young people concluded very decisively that media exposure to alcohol use 'increases the likelihood that adolescents will start to use alcohol, and to drink more if they are already using alcohol'. They found a 'dose–response relationship in all studies': i.e. the more media exposure, the more alcohol they (mis)used.[6]

A study published in the medical journal *Pediatrics* monitored adolescents aged ten to sixteen who had never drunk alcohol. The researchers had the pleasure of analysing and coding 398 films on the basis of alcohol use, defined as real or implied use of an alcoholic drink by any character in the film, including buying alcohol or occasions where alcohol was clearly in their possession (e.g. characters sitting at dinner with filled wine glasses), but actual alcohol use was not shown. The study reported that 'Movie exposure and having a television in the bedroom are both independent predictors of onset of problematic alcohol use ... Media restrictions could play a role in prevention.'[7] A related study found that alcohol presence in films was associated with the onset of teenage binge drinking. And yet again, 'there was a dose–response [relationship] for exposure to movie and alcohol use'.[8]

In 2010, researchers at Dartmouth Medical School found that among ten- to fourteen-year-olds whose parents let them watch R-rated (in the US this refers to 'Restricted' for those under 17) movies 'all the time', almost a quarter had tried a drink without their parents' knowledge. That compares with barely 3 per cent who tried a drink among those who were 'never allowed' to watch R-rated movies. The study, involving 3577 children who had never drunk alcohol, controlled for parenting style and still found 'the movie effect is over and above that effect'. The study gauged the children's perception of their parents' responsiveness and ability to set and enforce limits and even with such factors considered, the team still found a link between exposure to R-rated movies (90 per cent of which showed characters drinking alcohol) and the likelihood of early drinking of spirits. 'We think seeing the adult content actually changes their personality,' the researchers said, adding that PG13 movies, as well as many television programmes, also frequently portray drinking and other adult situations.[9]

Other research suggests that children who see R-rated movies become more prone to 'sensation seeking' and risk taking. Children and adults naturally differ in their inherent drive to seek sensation, but this can be artificially enhanced. A study of 6255 children published in *Prevention Science* in 2010 found that watching R-rated movies actually affected the level of sensation seeking among adolescents. It found that these movies not only portray scenes of alcohol use that prompt adolescents to drink, but that they also 'jack up the sensation-seeking tendency, which makes adolescents more prone to engage in all sorts of risky behaviours'. The researchers worry particularly about the effects of viewing cinematic misbehaviour on those adolescents who previously did *not* show signs of sensation seeking. High sensation seekers are already at high risk for use of alcohol, and watching a lot of R-rated movies raises their risk only a little. However, for low sensation seekers, 'R-rated movies make a big difference. In fact, exposure to R-rated movies can make a low sensation-seeking adolescent drink like a high sensation-seeking adolescent.' The message to parents is clear: 'Take the movie ratings literally,' says James Sargent, a professor of paediatrics at Dartmouth Medical School. 'Under seventeens should not be permitted to see R-rated movies.'[10]

The implications are significant because a growing amount of research shows that the younger children are when they begin to drink alcohol, the greater the risk that they will have a future problem (see Chapter 3). Starting to drink earlier is strongly linked to a fourfold increase in the risk of adult alcoholism.[11]

The influence of media on alcohol consumption in children and teenagers should be taken seriously, as a parallel 'media mechanism' is being observed in other young health behaviours. For example, strong links are now being found between screen media and sex. The suggestion is that screen media is hothousing young libidos and adolescent intercourse. A study published in *Pediatrics* in 2004 does what it says on the label: 'Watching Sex on Television Predicts Adolescent Initiation of Sexual Behavior' and found 'exposure to TV that included only talk about sex was associated with the same risks as exposure to TV that depicted sexual behavior.' The scientists involved identified an 'intercourse effect' among adolescents who either watched shows which only talked about sex or shows which had sex scenes: '... twelve-year-olds who watched the highest levels of this content among youths their age appeared much like youths two to three years older ... The magnitude of these results are such that a moderate shift in the average sexual content of adolescent TV viewing could have substantial effects on sexual behaviour at the population level.'[12]

And the effects reverberate beyond orgasm. In 2008, another paediatric study found that 'exposure to sexual content on television predicted teen pregnancy'. Teenagers (twelve to seventeen years of age, monitored up to fifteen to twenty years of age) who watch a lot of television featuring flirting, kissing, discussion of sex and sex scenes are twice as likely than their peers to get pregnant or get a partner pregnant.[13]

Obviously, underage drinking and underage pregnancy occurred before screen media were invented, but they do seem to have amplified these and other health behaviours significantly. This is especially so at a time when there are huge social changes afoot affecting children's lives, fundamental reasons why more and more young people at increasingly younger ages are turning to the bottle, and why binge drinking is on the rise. At one time, experimentation with drink was a rite of passage, but it has gone beyond that now.

## Who's In Da House?

The UK has the highest incidence of single motherhood in Europe. Studies of family structure around the world have found that children and young people who live with both biological parents are significantly less likely to use substances, or to report problems with their use, than those who do not.[14]

Two-parent households appear protective: children who are raised in single-family households are roughly twice as likely to experience alcohol-related problems such as alcohol abuse and alcohol dependency as compared with children who are raised by both parents in the same household. Children growing up without their biological father present are more likely to abuse drugs and alcohol as teenagers than those living with both biological parents. A typical study conclusion sums it up: 'The results obtained underscore the significance of the father as a key figure in the transmission of values and as a deterrent to certain behaviours. The results point to adolescents from fatherless homes, especially boys, as being at risk for problems.'[15]

In the west of Scotland, eighteen-year-old girls from lone-parent households were found to be twice as likely to drink heavily as those from intact two-birth-parent homes (17.6 per cent compared to 9.2 per cent). This finding holds even after controlling for poverty.[16] Another study of British sixteen-year-olds found that those from lone-parent households were 40 per cent *more* likely to drink.[17]

Some of the associations with family structure could be accounted for by differing levels of time spent in joint activities. It seems that the effect of lone parenthood on children's drinking is partly the result of one parent simply having less time than two parents to spend interacting with their child. And simple arithmetic suggests that four eyeballs are greater than two: one parent is not likely to be able to monitor and discipline a child as effectively as two.

A study in 2010 by the British government – 'Children, Young People and Alcohol' – considered family structure and parenting style and found that 'single parents (23 per cent) ... were more likely than average to say they have no fixed rules: this compares with 7 per cent on average'.[18] While a study published in the *Lancet* involving 986,342 children found that those from single-parent households were almost two and a half times more likely to suffer alcohol-related disease. And it isn't simply the result of poverty: 'Even when a wide range of demographic and socioeconomic circumstances are included ... children of single parents still have increased risks of mortality, severe morbidity, and injury.'[19]

And feminists aren't likely to enjoy reading the findings of a study in the *European Journal of Epidemiology*: 'Mother being a housewife was a protective factor [in] alcohol and illicit drug use in adolescents.'[20]

## Family Engagement

It isn't, of course, a simple case of family structure; it is also the quality and nature of relationships within that family structure (whether one- or two-parent) that influences the risk of alcohol use and abuse in children. For example, children who continue to see and have a good relationship with a non-resident father are less likely to have alcohol problems, while those who don't are more likely to have such problems. This is hardly surprising as we adults are also more likely to develop an alcohol problem if we have poor marital relationships and/or don't have enough contact with one another. The social monitor George Gallup Jr remarked that alcoholism was becoming a widespread problem because of the weakening of family ties, high mobility and lack of communication between parents and children.[21] Loneliness appears to be a source of vulnerability to alcohol problems.[22] In 'Loneliness and alcohol abuse: A review of evidences of an interplay', researchers concluded that 'loneliness may be significant at all stages in the course of alcoholism'.[23] While another found that loneliness was related to frequency of intoxication, binge drinking, and 'drink-tossing behaviours'.[24]

But loneliness isn't confined to adults, and a report from the NSPCC in 2010 found that more children are reporting they are lonely than in previous years; a detailed breakdown of calls made to their ChildLine in the previous five years showed that although, overall, the number of calls from children and teenagers had risen by just 10 per cent, calls about *loneliness* had nearly tripled. Among boys, the number of calls about loneliness was more than five times higher than it had been in 2004. NSPCC head of child-protection awareness, Christopher Cloke, said: 'Loneliness has always been a part of some children's lives, but it is deeply worrying that more children are contacting us about this.'[25]

## Well Connected?

New social technologies do not cause child alcoholism but they can provide a backdrop of social and emotional disengagement which lends itself to alcohol abuse. Drinking numbs the feelings of unhappy children.

We are constantly told that with so much modern communication technology and children's ability to understand how to use it, they've never been so connected to one another. Yet in explaining the role of technology in the rise in loneliness, the Mental Health Foundation's report 'The Lonely Society?' points out: 'The Internet has changed the way people communicate, but some experts argue that social networking sites like Facebook and Twitter undermine social skills and the ability to read body language ... technology doesn't provide the physical contact that benefits wellbeing. Cognitive function improves when a relationship is physical, as well as intellectual, because of the chemical process that takes place during face-to-face communication.'[26]

The rapid proliferation of electronic media is now making private space available in almost every sphere of the individual's life. Yet this is now the most significant contributing factor to society's growing physical estrangement. Whether in or out of the home, more people of *all* ages in the UK are physically and socially disengaged from the people around them because they are wearing earphones, talking or texting on a mobile telephone or using a laptop, iPhone or BlackBerry. Eye and ear contact between people of all ages and relationships is declining.

Children are now experiencing less social interaction and have fewer social connections during key stages of their physiological, emotional and social development.[27]

Time that was previously spent interacting socially is increasingly been displaced by the virtual variety. In 2008, an editorial of the *Journal of the Royal Society of Medicine* made the timely point that social networking 'encourages us to ignore the social networks that form in our non-virtual communities ... the time we spend socialising electronically separates us from our physical networks'.[28] But why precisely should physicians be concerned about these changes in people's *actual* contact and interaction with one another?

While many of us think of social networking as something done mostly by late teens and twenty-somethings, children a quarter of that age are now indulging. By age five, at least 25 per cent of children in Britain own their own laptop or computer and most have direct broadband Internet access. British children are watching more TV, while at the same time, Internet use is growing at a much faster rate and 'social networking is younger than ever ... the main reason to use the Internet'.[29]

Young children can now join the website 'MyCBBC' which the BBC describes as being 'about trying to develop their Internet skills and social networking'. And while we adults may have heard of 'Second Life', there is now a pre-school version – 'MyCBeebies' – where pre-schoolers are encouraged to 'create your own CBeebies Me: a personal avatar that represents you'. The Controller of BBC Children's TV says that this will 'help a child understand itself and its place in the world'. *The Times* recently published their 'Top seven social networking sites for kids'.

Yet, as children increasingly make virtual 'friends' online, they are making fewer friends in real life and spending far less time with the few friends they have. They are also spending less time looking at or talking to their own parents. In 2008, the Children's Society reported that television alone is displacing the parental role, eclipsing 'by a factor of five or ten the time parents spend actively engaging with children'.[30]

Even when they leave home for university, there has been a dramatic drop in the number of hours British students spend socialising.[31] Studies at Stanford University have led to 'a "displacement" theory of Internet use – time online is largely an asocial activity that competes with, rather than complements, face-to-face social time'. The researchers describe a 'hydraulic relationship' between the time spent on the Internet and the time we spend with our family and friends. And they use rather direct language: 'In short, no matter how time online is measured and no matter which type of social activity is considered, time spent on the Internet reduces time spent in face-to-face relationships.' On average 'an hour on the Internet reduces face-to-face time with family by close to twenty-four minutes'. And at weekends 'this means that for every minute spent online, there is a corresponding 0.48 seconds less spent with family members ... Time spent on the Internet at home has a strong, significant negative influence on time spent with family members.'[32]

We should all do the arithmetic – there are still only twenty-four hours in a day.

## Empathy

Yet another role for alcohol is created by the fact that when teenagers do have the time and friends to turn to for support, those peers are now likely to be less understanding than they used to be. Today's university students are not as empathetic as those of the 1980s and 1990s. A University of Michigan and University of Rochester Department of Psychiatry study undertook a meta-analysis, combining the results of seventy-two different studies on empathy conducted between 1979 and 2009 among almost 14,000 university students. And there was little flattery for the young: 'We found the biggest drop in empathy after the year 2000. College kids today are about 40 per cent lower in empathy than their counterparts of twenty or thirty years ago, as measured by standard tests of this personality trait.'

The researchers sought a second opinion too, and in a related but separate analysis they found changes in other people's kindness and helpfulness over a similar time period in nationally representative samples of Americans. Many people see the current generation of university students – 'Generation Me' – as 'one of the most self-centred, narcissistic, competitive, confident and individualistic in recent history ... It's not surprising that this growing emphasis on the self is accompanied by a corresponding devaluation of others.'[33]

In looking for the culprit who stole the empathy, the fingerprints of electronic media were, yet again, everywhere. The investigators believe that the sheer increase in child and adolescent exposure to media during this time could be one very important factor. They noted that compared to thirty years ago, the average American is exposed now to three times as much non-work-related information. And, in addition to the amount, the content is also having an effect. 'In terms of media content, this generation of college students grew up with video games, and a growing body of research, including work done by my colleagues at Michigan, is establishing that exposure to violent media numbs people to the pain of others.'[34] This is also supported by neurological and electrical skin-conductance research finding that the more violent the content of a video or the more time boys spent watching, the less emotional reaction to seeing violence.

The University of Michigan study concluded that the rise of social media may also play a role in the drop in empathy, 'The ease of having "friends" online might make people more likely to just tune out when they don't feel like responding to others' problems, a behavior that could carry over offline.'

Add to all of this the hypercompetitive atmosphere and inflated expectations of success, borne of celebrity 'reality' shows, and you have a social environment that works against slowing down and listening to someone who needs a bit of sympathy. 'College students today may be so busy worrying about themselves and their own issues that they don't have time to spend empathising with others, or at least perceive such time to be limited.'[35]

With this in mind, a study published in *Proceedings of the*

*National Academy of Sciences* (2009) examined the brain function and development which underlie abstract qualities such as empathy. The scientists drew specific attention to the effects of electronic media: 'The rapidity and parallel processing of attention requiring information, which hallmark the digital age, might reduce the frequency of full experience of such emotions, with potentially negative consequences.' One of the authors explained the possible interference with this process by the speed of today's media: 'For some kinds of thought, especially moral decision-making about other people's social and psychological situations, we need to allow for adequate time and reflection. If things are happening too fast, you may not ever fully experience emotions about other people's psychological states.'[36]

The damage done by displacing key periods of emotional and social development with time in front of a screen doesn't have the sense of dramatic risk that Internet paedophilia does. However, increasing time spent in a virtual world displaces vital development time spent experiencing real socialising, learning to interpret, respond to and cope with the nuances of real emotion, relationships, disappointments and disagreement – the human condition. Children need to spend vital time experiencing *real* social interactions with *real* children and young adults face to face.

Social technologies have had other unintended consequences, allowing far greater social comparison from the comfort of your bedroom. And so children's lives have become more competitive and feelings of failure are lived out in the virtual public view.

Drinking also numbs the fear of rejection and, in a world where our children are not as used to expressing themselves face to face as perhaps they once were, it fills a social confidence gap too. A colleague recently told me about a teenage pupil who told him how she had to 'pre-load' with drinks before going to any party, explaining, 'I can't approach and talk to a boy unless I'm drunk.'

## Family Disengagement

Referring to his resignation as US Secretary of Labor, in 2001, Robert Reich cited his concern that long working hours harmed his relationship with his sons:

> I did have to leave the best job that I'd ever had and probably ever will have and I came home ... They are exactly like clam shells. They are tightly shut and occasionally, just occasionally, when you least expect it, those clam shells open and you see inside this very soft and beautiful and very vulnerable interior. Then the clam shell shuts tight again and you don't see it and you don't know when, if ever, it will open. But it will open at a very unexpected time and in a very unexpected way, and if you're not there when it opens you might as well be on the moon.[37]

Many parents worry justifiably that work demands diminish their opportunities for interaction with their own children and that, over time, this may compromise their future relationship. The number of dual-earner households with children has jumped in recent decades and a spokeswoman for one research body has described it as, 'Two people. Three full-time jobs ... I call it the new math.'[38]

For a number of years the Center on the Everyday Lives of Families (CELF) at the University of California conducted a detailed database of middle-class families. They were filmed and computer monitored for weeks and all social interactions between family members were analysed. One of the many papers published analysed interaction among dual-income family members after 3 p.m. when children came home from school. And there were some *very* modern findings.

In measuring things such as 'physical proximity in home spaces' they reported that 'family members seldom came together as a group. On average, all family members in the thirty families came together in a home space in only 14.5 per cent of the observation rounds. In contrast, individual family members were observed alone in a home space with far greater frequency – averaging 30–39 per cent of the observation rounds.' The most telling stat was that children were found alone in almost 35 per cent of observed cases.

They also measured the degree of child distraction, defined as not acknowledging their returning parent because they were 'otherwise engaged in activity (e.g. watching TV, playing video game, phone)'. And the findings become more insidiously chilling, as the number of parents who were ignored or unacknowledged on their return home 'comprised a substantial percentage of observed behavior. The high level of distraction encountered by fathers when they reunited with their children was particularly striking ... distraction was displayed by at least one child in the family in over two-thirds of the twenty-nine father–child reunions ... fathers were more likely to be the recipients of distraction from at least one child in the family (86 per cent) than were mothers (44 per cent).' In the bigger picture, from a cross-cultural perspective on how children normatively greet their parents and others, the scientists agree: 'These latter results are particularly noteworthy. Social scientists have long documented the near universality of positive behavior in the form of greetings when two or more people reunite after being apart for a period of time. Greetings recognise a person's arrival, status and display positive intentions that universally facilitate the transition into social interaction with another.'[39]

Furthermore, Britain has the lowest proportion of children in all of Europe who eat with their parents at the table.[40] But the good news is that parents, whether single or dual, can actively redress some of the disadvantages conferred by this, as well as those spelled out above, through simple means. A study at Columbia University reports that having at least one parent eat dinner with their child regularly was found to prevent depression, anxiety and substance abuse in children. Research at the National Center on Addiction and Substance Abuse at Columbia University 'consistently finds that the more often children eat dinner with their families, the less likely they are to smoke, drink or use drugs'. The Chairman of the National Center has even gone so far as to state: 'One of the simplest and

## Three

# Into the Mouths of Babes

Peter Mayle's *A Year in Provence* was a catalyst in leading many couples to aspire to the good life across the Channel. Dr Michel Craplet, a French psychiatrist and chairman of the European Alcohol Policy Alliance, is bemused, saying, 'Many people who come here for holidays think France is a paradise.' British middle-class Francophiles continue to gaze across the Channel for cultural guidance in matters of food, film, fashion, cars ... and wine. And this extends to an admiration of the seemingly inclusive and intuitive way in which the French gradually introduce their children to alcohol.

## French Lessons

Visitors to France who admire the manners and civilised behaviour common among French children might be interested to know that French parents also cultivate this in a most 'intuitive' way. French parents are the most heavy-handed in Europe: nearly nine out of ten adults have spanked their children. In addition, nearly a quarter of French parents have slapped their children across the face and 10 per cent admit to punishing their child with a 'martinet' – a small whip resembling a cat o' nine tails. In fact, next time you pop over to Boulogne for a case or two of Pinot noir you can pick one up, as they are apparently still on sale in the pet section of French supermarkets; it is generally believed that a large share of those bought are intended for use on children, not pets. This aspect of French culture looks likely to continue with future generations; when asked how they planned to discipline their own children when they become parents, 64 per cent of French children responded 'the same'.[1]

## Frog-marched to *la bouteille*

I never cease to be amazed at the sycophantic adoration of the British middle classes for the way in which the French drink. Yes, the French lead the world in wine and have good reason to persuade vintners in Leicestershire to stick to what they do best – making pork pies, not cultivating grapes. But what I saw in Paris, when I lived there for a year after finishing high school (that is children drinking table wine in water glasses at meals), tallies perfectly with the European Union health statistics, clearly demonstrating to us that the French have nothing whatsoever to teach us about alcohol.

Dr Craplet encapsulates the French paradox nicely: 'The problem is that every French person is a lobbyist for wine. It's in the head, in the culture. We don't need the alcohol lobby here because we view wine passionately. We have a conflict between the figures about drinking that prove our mortality and morbidity and the positive symbolic value it has.'

Although parents may delude themselves into believing that they are giving their children a more responsible cosmopolitan and sophisticated approach to alcohol, it is worth remembering that France's death rate from cirrhosis of the liver was until very recently actually twice that of England and Wales.[2]

British liver specialists have started to complain aggressively about this misperception that it is British binge drinking that causes more liver damage. A team at the University of Southampton Medical School has published a study emphasising how 'The link between daily or near-daily drinking and the development of liver disease is important ... Conclusions: Increases in UK liver deaths are a result of daily or near-daily heavy drinking, not episodic or binge drinking, and this regular drinking pattern is often discernible at an early age ... only a minority of patients with cirrhosis or progressive cirrhosis have evidence of severe alcohol dependency and many have no evidence of dependency at all – they are controlled, heavy social drinkers.'[3]

The point being, that while 'binge drinking' may be one particularly antisocial British route to some types of health damage, even the slow, stylish continental way will get you there.

## Doing the Continental

A more nuanced and qualified Francophobia is emerging very meekly in some of the right places. A study commissioned in 2008 by the UK's Department for Children, Schools and Families cited the myth of the benefits of the continental approach to introducing children gradually to alcohol as a 'huge obstacle' to overcome in protecting British children. 'The misperceptions are firmly based on opinion rather than from health statistics about mainland Europe. Parents … are searching for any logic that helps them maintain their own drinking whilst protecting their children.'

Another major government study previously referred to in Chapter 1, 'Children, Young People and Alcohol', published in 2010, divided alcohol parenting style into categories including 'the Educating Liberals', 'Strong Rejectors', 'Risk-reducing Supervisors', and the plain old 'Stressed and Concerned'. Young drinking was cited as a particularly 'middle-class problem', and the study revealed that affluent, liberal families are most likely to follow the 'continental' practice of letting children have a small amount of alcohol. 'In general, white parents, heavier drinkers and AB parents [the richest social group] were more likely to think that it is acceptable for children to start drinking at a younger age.' They found that 70 per cent of parents 'think it's safer to introduce their child to alcohol gradually, like they do in Europe':

> Parents who said that they drank more than the recommended guidelines were also more likely to agree that they think it's

safer to introduce their child to alcohol gradually (85 per cent) than those who said they drank within the recommended guidelines (68 per cent). Interestingly, Black (23 per cent) or Asian parents (18 per cent) were less likely to agree with this 'continental model' than White parents (77 per cent).[4]

## 'Do As I Do'

Despite the beastly French, in the recent trend towards so-called positive discipline, parents have been encouraged to be friends with their children, to provide them with choices in many matters and to negotiate more. To enhance a child's wellbeing, 'you've got to accentuate the positive, eliminate the negative'. Unsurprisingly, these values and expectations have travelled from the home to the school and society in general, as we adults in general have been encouraged to be more friendly and co-operative with children.

This is part of a modern belief that we should negotiate with our children because it seems more democratic – and it's also just plain nicer to endear ourselves to them, as opposed to challenging and upsetting them. It also reflects a misplaced concern that to demand our children's compliance, through compulsion if necessary, will in some way be counterproductive and even harmful, quashing their character and preventing them from expressing themselves. Yet the result hasn't been what we expected. Although paved with the best of intentions, this permissive approach to parenting has actually been highly counterproductive. Without clear boundaries and clear figures of authority, not only do our children develop a sense of entitlement and self-centredness, they are also less happy, secure and socially viable.

We have actually retreated from parenting. Many of us seem confused: unable to distinguish confidently between being authoritative and authoritarian, we have chosen what appears to be the safer option. And many teachers, policemen, doctors – figures of authority – have gone to great lengths to obscure obvious signs of any pecking order and control. This loosening up of overt hierarchy and power relations may seem cosy and kind, but it has helped to undermine the authority that children desperately need.

In line with this, I increasingly see apologetic messaging from the authorities to children concerning alcohol. For example, a 'Please, would you mind ...?' approach to exerting authority over children and the young popped up at my local supermarket checkout counter, where a sign reads: 'Please don't be offended if we ask for proof of age when you buy alcohol.' And I've just taken the longest flight on earth – almost 12,000 miles from London to New Zealand – only to find a Southern hemispheric variation on this theme: 'Please don't be offended if we ask for proof of age when you buy alcohol. *Please take it as a compliment!*'

Parents feel unconfident and uncomfortable, not entitled to say, 'Do as I say, not as I do,' rather than feeling entitled to say, 'Your father and I are adults and can do a lot of things that you can't do simply because we are adults and you are not. We are also your parents and we are responsible for what goes on in this house. For example, if the mood takes us we can slither upstairs after dinner and have sex in the doggy position; in fact, after you're asleep we can do it on the sofa or kitchen table. You, by the way, cannot do this.'

There is no reason why children should not learn about the fundamental differences in rights, power and entitlement between adults/parents and children – if anything, this is the defining difference between them. Parents have been led to believe that if they drink at home, this is in some way a contradiction which sends mixed messages, unless they allow their children to have some alcohol too.

And where other families have different alcohol rules for their children, modern parents have also developed an aversion to saying, 'Do as I say, not as other children do,' or, 'Do as I say, not as other parents say.' This would mean learning that people can have or do things that we don't have or can't do. On any street at any given time, children see other children who have things or are doing things that they don't have or can't do. The neighbour's children aren't allowed to eat hamburgers in their house, our children aren't allowed to own PlayStations in our house . . . Children must understand that different families, even neighbours or relatives, have different values and rules and that alcohol is merely one of these.

There is a fear that to forbid our children to drink will invite a backlash, bringing on the opposite of what we intended in that they will actually drink more as a result, and may be more likely to become alcoholic. And so parents buy alcohol for their underage children's birthday parties, allow them to go to pubs when they are underage and serve them wine at the dinner table early and moderately to wean them on to alcohol in a controlled, responsible environment. (In fact, incredibly, the legal age in the UK for drinking alcohol in the home is five!)

Parents want to prevent their children from doing things behind the bike shed as children did in the repressive, authoritarian 'bad old' days.

## Forbidden Fruit?

The zeitgeist has been one of transparency and openness. Better out than in. There is also a misconception that to deny adolescents *some* alcohol elevates it to the status of a forbidden fruit. The best way to demystify alcohol, it is thought, is to let them drink it. All of this is intended to cultivate 'sensible drinking'. You can see, however, that this basic change in our relationship with our children has implications not just for alcohol but for many other unrelated aspects of their behaviour and wellbeing.

A growing number of studies confirm common sense: conveying your values, alcohol rules and boundaries to your children is more likely to *prevent* your child from binge drinking or developing an alcohol problem. In particular, *parental disapproval is good for child sobriety.* For those who find it hard to believe, here is the conclusion of a study published in the *Journal of Studies on Alcohol and Drugs*: 'Authoritative parenting appears to have both direct and indirect associations with the risk of heavy drinking among adolescents. Authoritative parenting, where monitoring and support are above average, might help deter adolescents from heavy alcohol use, even when adolescents have friends who drink.'[5]

This is found on opposite sides of the world. In Australia, a team of researchers looking for 'Parenting factors associated with reduced adolescent alcohol use' reported that buying alcohol for underage teenagers to take to parties or allowing them to drink at home can increase the risk of problems later in life. 'We found that delayed alcohol initiation was predicted by: parental modelling, limiting availability of alcohol to the child ... Reduced levels of later drinking by adolescents were predicted by: parental modelling, limiting availability of alcohol to the child, disapproval of adolescent drinking, general discipline, parental monitoring.'[6] And the same things show up in New Zealand. In answer to the question 'How to reduce alcohol-related problems in adolescents: what can parents do and what can the government do?' the National Addiction Centre, the University of Otago, found that supplying alcohol to young people under supervision backfired as a tactic aimed at reducing harm. 'Many parents consider this is the best way to prevent negative alcohol outcomes in their children, that is by allowing drinking at home and directly supplying them with small amounts of alcohol when they go out to parties. The normalisation of drinking alcohol is aimed at lessening the "big deal" of adolescent initiation rites involving alcohol. However, the evidence points in the opposite direction, that normalisation of alcohol increases the risk of harm.' Lower levels of alcohol use later in life were noted in those whose parents monitored their activities and knew their friends, and the same applied to parental disapproval of adolescent drinking. One of the best things parents can do to minimise alcohol-related problems in their adolescents is, it was found, to expose them to healthy alcohol behaviour, rather than drunkenness.[7]

The Society for Prevention Research is a prominent international body dedicated to 'advancing scientific investigation on the etiology and prevention' of social, physical and mental health problems. One of those preventions is an alcohol problem. At their annual meeting in 2010, they too heard similar findings from Caitlin Abar and colleagues of the Prevention Research and Methodology Center at Pennsylvania State University. In their symposium 'Myth of the forbidden fruit', she suggested that parents practise a zero-tolerance policy in the home and said that there is no scientific basis to the common belief that prohibiting alcohol turns it into a 'forbidden fruit' and encourages abuse. Abar studied three hundred first-year university students and compared their drinking habits to their parents' attitudes towards alcohol. Those students whose parents never allowed them to drink – about half of the group – were significantly less likely to drink heavily at university, regardless of gender.

Moreover, 'The greater number of drinks that a parent had set as a limit for the teenagers, the more often they drank and got drunk in college,' said Abar. Whether the parents themselves drank, on the other hand, had little effect on predicting their children's behaviours.[8]

A previous study in 2004, on 'Adults' approval and adolescents' alcohol use', by Wake Forest University Baptist Medical Center in North Carolina, showed that teenagers who received alcohol from their parents for parties were up to three times more likely to binge drink within a month.[9]

Related research has examined parenting and alcohol during the transition period between the end of secondary school and the first year at university. The researchers looked at three aspects of drinking: weekly alcohol use, 'heavy episodic' (binge) drinking and alcohol-related problems. They found that the degree of 'parental permissiveness' towards drinking and the student's intention to be involved with fraternities or sororities 'predicted the transition to use and consequence status for all three outcomes and for increases in alcohol use and consequences … Our findings indicate the importance of the parental context (e.g. parental permissiveness of drinking).'[10] A study published in *Developmental Psychobiology* looked at the importance of 'sustained parenting' at the late stages of finishing secondary school and leaving home in reducing drinking at university in first-year students. The study found 'excessive drinking in college is a continuation of high-school drinking tendencies'. The researchers concluded that to prevent alcohol misuse in their children's transition to university, parents should remain vigilant and intolerant towards their teenagers drinking. 'The empirical evidence from this study suggests that sustained parental efforts have a beneficial effect on reducing high-risk drinking and preventing harm even at this late stage of late adolescent/early-adult development.'[11]

Parenting style is also important in 'inoculating' children who go to schools where drinking is prevalent among other pupils. An American study involving twelve- to fifteen-year-olds in forty-nine metropolitan schools found that students who receive 'suboptimal' parenting may benefit from increased support to 'deter them from early alcohol use, especially in high-risk schools'.[12]

This theme seems to be found across continents, income, class and race. For example, a similar effect has been found among low-income black and Hispanic students. A 2007 study of 1388 children by Kelli Komro, a professor at the University of Florida's College of Medicine, found that schoolchildren who were allowed to drink alcohol in the home by their parents at age twelve were up to three times more likely to get drunk and almost twice as likely to drink heavily (five or more drinks) at ages twelve to fourteen.[13]

So while parents who try to teach responsible drinking by letting their teenagers have alcohol at home may be well intentioned, they may also be wrong. In a study of 428 Dutch families in 2010, researchers found that the more teenagers were allowed to drink at home, the more they drank outside the home as well. And teenagers who drank under their parents' watch or on their own had an elevated risk of developing alcohol-related problems, including trouble with schoolwork, missed school days and getting into fights, among other issues. In addition, teenagers who drank more often, whether in or out of the home, tended to score higher on a measure of problem drinking two years later. Parental supervision of adolescents' alcohol use did not have a positive effect at all. The suggestion is that teenage drinking begets more drinking and, in some cases, alcohol problems regardless of where and with whom they drink. Their findings, say the researchers, throw into question the advice of some experts who recommend that parents drink with their teenage children to teach them how to drink responsibly with the aim of limiting their drinking outside of the home. This advice is common in the Netherlands, where the study was conducted, but it is based more on experts' reasoning than on scientific evidence, according to Dr Haske van der Vorst, the lead researcher on the study.

Based on this and earlier studies, van der Vorst says, 'I would advise parents to prohibit their child from drinking, in any setting or on any occasion ... If parents want to reduce the risk that their child will become a heavy drinker or problem drinker in adolescence, they should try to postpone the age at which their child starts drinking.'[14]

Another way of looking at this issue is to see what happens when parental attitudes and rules on alcohol are changed from permissive to prohibitive. In the cold, northern climate of Scandinavia, researchers at Orebro University in Sweden published clear experimental evidence in favour of prohibiting alcohol in the home. They designed a no-drinking intervention programme for thirteen- to sixteen-year-olds that cut teenage drunkenness by 35 per cent.[15]

## Ambient Parenting – Passive Drinking

Beyond overt parental approval/disapproval of your child drinking is the more ambient effect of a mother's general belief that her child will or won't drink – a self-fulfilling prophecy effect. And a study in 2008 on 'Mothers' Self-fulfilling Effects on Their Children's Alcohol Use' found precisely that: 'There was a significant indirect effect of mothers' beliefs on children's alcohol use.'

The study of 775 mothers and their adolescent children found strong evidence that a mother's beliefs regarding her child's likelihood of using alcohol altered her child's 'self-view' in either a positive or negative direction. The child then validated that new self-view by acting consistently with it later on. 'The more acceptable teens believed adolescent alcohol use was, the more alcohol they tended to drink themselves.' In these adolescents, alcohol use at age thirteen had doubled by age fourteen and had tripled by age fifteen, simply because of their mothers' cumulative self-fulfilling effects. So when mothers overestimated their teenagers' future use of alcohol, the teenagers developed the self-view that they were likely to drink alcohol in the future, which ultimately led them to drink more.[16] This sense of inevitability is pervasive in countries such as Britain, where parents and supposedly neutral 'alcohol-awareness' organisations subscribe to a throwing-your-hands-up-and-shrugging-your-shoulders position – 'Well, what can you do?'

Children pick up subliminally on a variety of things related to a parent's attitude to alcohol in ways that seem extraordinary. They have now been found to connect alcohol odours with their mother's emotions. A study of children aged five to eight asked the children's mothers to complete a questionnaire about their drinking habits, including their reasons for drinking; thirty-five were classified as 'Escape drinkers', based on their having indicated at least two 'escape' reasons for drinking (these included: it helps to relax, need when tense and nervous, helps to cheer up when in a bad mood, helps to forget worries and helps to forget everything).

Because information from odour travels directly to areas of the brain that process non-verbal aspects of emotion and memory, studying children's responses to odours provides insights into their emotional worlds. Like adults, children are not very good at identifying odours, but they are good at telling us whether they like an odour or not. This study demonstrated that 'whether they like the odour of beer depends not just on how often their mother drinks, but on why she drinks'. When asked to choose between the odour of beer and rotten eggs, children of the Escape drinkers were more likely than children of Non-escape drinkers to choose the rotten eggs.

Questionnaires also revealed that Escape drinkers drink more often than Non-escape drinkers. Because of this, children of Escape drinkers were exposed to alcohol odours more often. These children also experienced alcohol in a different emotional context.

Even before their first taste, young children are learning about alcohol and about why their parents drink. They do this by seeing people drink and hearing them talk about it. The lead researcher said, 'Our findings show that children are also processing the smell of alcohol with the emotional reasons their mothers, and perhaps fathers, drink.'[17]

## Pre-birth Parenting Style

The good and bad news, depending on how one wants to view it, is that we influence our adolescents' attitude to alcohol long before they're even born. Epidemiological studies are finding that foetal alcohol exposure is a strong predictor of adolescent craving for and abuse of alcohol.[18] A new generation of prenatal studies is clearly finding that the foetus can detect and later remember the alcohol it was exposed to in the womb. And years later this pre-birth exposure increases alcohol consumption in the adolescent 'by making it smell and taste better' than would normally be the case. One of a number of studies on rats published in the *Proceedings of the National Academy of Sciences*, reveals that exposure to alcohol in the womb alters gene expression in the developing foetus, thereby changing 'the development of the smell and taste systems so that the normally aversive odour and flavour of ethanol became more acceptable, thereby enhancing intake'. The scientists concluded, 'Here, we describe an epigenetic mechanism by which maternal patterns of drug use can be transferred to offspring.'[19]

Even the degree of influence that binge-drinking peers have on our adolescent child seems to be partly determined by their exposure to alcohol while in the womb. Rats whose mothers were fed alcohol during pregnancy were found to be more attracted to the smell of liquor during puberty. Researchers have shown that rats exposed during gestation find the smell of alcohol on another rat's breath during adolescence more attractive than animals with no prior foetal exposure, thereby 'promoting interactions with intoxicated peers'. In this study the authors found that those rats not exposed to alcohol in the womb were significantly less likely to follow an intoxicated peer than those that were.[20]

A statement by the Neuroscience and Physiology Laboratory of the Upstate Medical University in New York sums it up nicely: 'Human studies point to a causal relationship between foetal alcohol exposure and adolescent ethanol [alcohol] abuse. Foetal exposure is, perhaps, the best predictor of abuse in this "at risk" age group, and there is an inverse correlation between the age of first experience and continued abuse.'[21]

Foetal exposure to even low levels of alcohol has other unexpected effects on parenting, including the actual bond between a mother and her infant. British researchers found that a third of the mothers they studied in 2009 drank during pregnancy; of these, just under half drank only one or two units 'monthly or less', while a similar proportion drank 'two to four times a month' and a small minority drank 'two or three times a week'. The effects of alcohol use on mother–infant attachment were examined and the three female researchers didn't mince their words: 'Women who drank during pregnancy had significantly lower attachment scores than abstainers. Drinking during pregnancy has negative outcomes for women. Alcohol use during pregnancy was associated with lower attachment scores, indicating the potential impact on mother–infant relationships.' This finding held true even while controlling for the mothers' existing levels of depression, anxiety and stress.[22] The weaker attachment between mother and infant may be due to the alcohol having affected the infant's ability to be responsive.

Previous research has found a relationship between prenatal alcohol use and infant behaviour. A study entitled 'Binge alcohol consumption by non-alcohol-dependent women during pregnancy affects child behaviour' found that 'children exposed in utero to maternal binge drinking displayed a greater degree of disinhibited behaviour and that this behaviour was associated with early drinking variables ... binge drinking in pregnancy does increase the likelihood of certain behavioural characteristics that might predispose these children to later behavioural dysfunction'.[23]

Other studies on women who are neither dependent on alcohol nor drink regularly during the second and third trimesters of pregnancy are now finding that having less than four drinks in a day on an occasional basis during pregnancy 'may increase risk for child mental-health problems in the absence of moderate daily levels of drinking. The main risks seem to relate to hyperactivity and inattention problems.'[24] And there's more. Drinking during pregnancy appears to be associated with conduct problems in children, independently of other risk factors. A twenty-five-year study of 4912 mothers, followed later on by 8621 of their offspring, found that for each additional day per week that mothers drank alcohol during pregnancy, their children had an increase in conduct problems. This association remained even after factoring in other variables, such as the mothers' drug use during pregnancy, education level or intellectual ability. The study included multiple children per mother, which allowed researchers to look at siblings who were exposed differently to alcohol prenatally because their mothers varied their drinking during different pregnancies. The study found that children more frequently exposed to alcohol during pregnancy had more conduct problems than their siblings who were exposed to less prenatal alcohol. 'The findings thus support a strong inference that prenatal alcohol exposure causes an increased risk of offspring conduct problems through environmental processes,' the author concluded.[25]

## Teaching Sensible Snorting

As we have seen, a myth persists that introducing children to alcohol earlier prevents heavy drinking and alcoholism later. While many believe that children benefit from the role modelling and restraint displayed at the family dinner table, they have not considered the biochemical, genetic and neurological processes at work. In fact, exposure to alcohol at an early age is *more* likely to increase the chances of a child developing alcohol use disorders. And we shouldn't be in the slightest bit surprised. Be it anything from nicotine and caffeine to cocaine – introducing our children to small amounts of almost any addictive substance is a fantastic way to ensure that they will be more likely to end up dependent on the substance.

A study in 2008 from the US government's National Institute on Alcohol Abuse and Alcoholism found that having a first drink of alcohol before the age of fifteen (not counting small tastes or sips) sharply increased the risk of alcohol-use disorders (abuse and dependence) that persist into adulthood. The researchers believe it is important to delay the 'age of first drink' as long as possible.

In this longitudinal study, children and young people were screened for the age they had their first drink and indications of alcohol dependence or alcohol abuse. They were then assessed again when they were older and 'the odds of developing alcohol dependence were significantly increased for individuals who initiated drinking before age fifteen or at ages fifteen to seventeen'. There was also a significant link between the age of first drink (AFD), whether before age fifteen or between the ages of fifteen and seventeen, and later incidence of alcohol abuse 'even after adjusting for all significant covariates'.

They then looked specifically at 'low-risk drinkers' and even found that in those who had their first drink before the age of eighteen the risk of developing alcohol dependence 'was far greater' than the comparable risk for the total population. One particular significant risk that increased in these 'low-risk' teens who had their first drink before eighteen was that they 'continued drinking despite physical/psychological problems caused by drinking', as opposed to those who had their first drink at eighteen or older.

The researchers' conclusions will hardly have the drinks lobby popping the champagne corks. They pointed out in very clear terms that there is an 'extremely robust' association between the age at which children have their first drink and the incidence of alcohol problems. And they expressed particular concern over the fact that this link between the age of first drink and later alcohol problems was apparent in 'such a resilient population, and that it did not diminish with age'. Spelling out the message for society, the researchers leave us with their final thoughts on the matter: 'Because alcohol-use disorders are so common and have such devastating consequences for affected individuals and their families, even an association of modest magnitude has major clinical and public health implications.'[26]

A robust link between the AFD and later alcohol-use disorders has been found in many previous studies,[27] but more recent studies have managed to control for other pre-existing risk factors.

Washington University School of Medicine recently studied 6257 identical versus non-identical twins and came to the same conclusion for the under-sixteens: 'Early age at first drink may facilitate the expression of genes associated with vulnerability to alcohol dependence symptoms. This is important to consider, not only from a public health standpoint, but also in future genomic studies of alcohol dependence.' Using the twin model, the researchers were able to tease out genetic influences, shared environmental influences and non-shared environmental factors that contribute to the risk of later developing alcohol-dependence problems. They assessed the twins when they were twenty-four to thirty-six years old and found that when twins started drinking earlier, genetic factors contributed greatly to their risk for alcohol dependence, at rates as high as 90 per cent in the youngest drinkers. For those who started drinking at older ages (over sixteen), genes explained much less, and environmental factors that make twins different from each other, such as unique life events, gained prominence. In short, the younger a person's AFD, the greater the risk for alcohol dependence and the more prominent the role played by genetic factors.

The researchers understandably preach the principle of precaution: 'Something about starting to drink at an early age puts young people at risk for later problems associated with drinking. We continue to investigate the mechanisms, but encouraging youth to delay their drinking debut may help ... we could hypothesise that exposure to early-onset drinking somehow modifies the developing brain.'[28]

Others have gone on to identify the actual mechanisms. I've left the technical language in the following passages so that any sceptics may get a feel for the profundity of this new generation of research which has concrete implications for our children's brains and development: 'Alcohol during adolescence selectively alters immediate and long-term behaviour and neurochemistry.' The scientists make it clear that 'early ethanol [alcohol] exposure induces long-term changes in responsivity to ethanol in adulthood. Exposure to moderate doses of ethanol during adolescence produced alterations in dopamine in the nucleus accumbens septi during adolescence and later in adulthood. Taken together, all of these data indicate that *the adolescent brain is sensitive to the impact of early ethanol exposure during this critical developmental period.*'[29] Cellular pathologists on the other hand, writing in the journal *Alcohol*, state: 'Experimental evidence suggests that early exposure to alcohol sensitises the neurocircuitry of addiction and affects chromatin remodelling, events that could induce abnormal plasticity in reward-related learning processes that contribute to adolescents' vulnerability to drug addiction.'[30] The bottom line? That alcohol seems to affect children's brains in very specific and increasingly identifiable ways.

The next chapter will look in more detail at the effects of alcohol on the developing brain. But one point worth considering now is that the young brain is very malleable and changes quickly in response to new influences; early exposure to alcohol may 'prime' the brain to enjoy alcohol by creating a link between it and pleasurable reward. The same is true with nicotine. It seems that by trying to help our children resist overindulgence with alcohol, we may inadvertently be switching on genes that affect a susceptibility to alcohol addiction.

The US Department of Health and Human Services National Institutes of Health has now clearly concluded: 'A person who begins drinking as a young teen is four times more likely to develop alcohol dependence than someone who waits until adulthood to use alcohol.'[31] The UK government's Chief Medical Officer made it explicitly clear in 2009 that children should not drink *any* alcohol until they are *at least* fifteen years old;[32] and in 2007 the American government's Surgeon General made it clear that no one should drink alcohol until they are at least twenty-one years old. This includes parents offering their twenty-year-olds a glass of wine at home.[33] Neither position is likely to be popular among teenagers and young voters, and governments may lose sin tax, so we have to ask ourselves why would such unpopular decisions have been made. Obviously, the governments are more than convinced that drinking alcohol earlier than this is far more likely to lead to alcohol problems, 'alcoholism' and harm to health later. (Note: the reasons for the six-year disparity – fifteen v. twenty-one years are political and cultural, not medical, and will be discussed in the final chapter.)

The organisation Alcohol Concern has called for parents who give alcohol to children under fifteen to be prosecuted.[34] Yet in Britain there is great inconsistency and contradiction between the medical evidence and position on the one hand and both parental and school positions on the other. For example, Britain's leading education newspaper, the *TES*, in a section called 'Analysis', offered tips on how schools should deal with alcohol, including: 'Encourage pupils to discuss why they drink and why they want to get drunk. Follow up with wine-tasting sessions (with parental permission and in accordance with legal guidelines) to encourage pupils to think of alcohol as something that can savoured.'[35]

## Gene–Environment Interactions

Our children's preference for and tolerance of alcohol is genetically influenced, as is the risk of them developing an alcohol use disorder. Predisposition to alcoholism runs in families. A child with an alcoholic parent is four to ten times more likely to develop alcoholism themselves. The extent to which a person's predisposition to alcoholism is inherited is thought to be about half: the heritability of predisposition to alcoholism, arrived at from studying nearly 10,000 twin pairs, is 50 per cent.[36] Thus, genetic and environmental factors are almost equally important in alcoholism risk.

And there are also both individual and racial genetic differences in alcohol-metabolising genes seen in Asian and Jewish populations – but, unsurprisingly, somewhat missing in action among Anglo-Saxons, Celts and northern Europeans – which provide protection against the development of heavy drinking and subsequent alcoholism. For example, approximately half of Japanese, Chinese and Koreans, together with other East Asians, have differences in four key genes. So when people with two particular types of these genes (ALDH2*2 and ADH2*2) drink even small amounts of alcohol they experience a very unpleasant reaction characterised by facial flushing, headache, low blood pressure, palpitations, abnormally rapid heartbeat, nausea and vomiting – ironically, many of the same attractive symptoms featured on display in most city centres on any Friday or Saturday night in today's Britain. One gene type is nearly completely protective against heavy drinking, while another is partially protective. It has been suggested that the relatively high frequency of one gene type (ADH2*2) might be implicated in the lower levels of alcohol consumption and increased sensitivity to alcohol observed among Jews.[37]

As mentioned earlier on, environmental influences, such as alcohol availability, parental attitudes and peer pressure, strongly affect if and when a child starts to drink, and frequent drinking in adolescence has been shown to independently increase the risk for alcoholism. Severe childhood stressors, especially emotional (harsh, inconsistent discipline, hostility, rejection), physical, and sexual abuse, have been associated with increased vulnerability to addiction. Childhood sexual abuse is associated with a fourfold increase in the lifetime prevalence of alcoholism and other drugs of abuse in women.[38]

But not all children who experience adverse events subsequently develop psychopathology (psychological problems) predictive of alcoholism or take up drinking. So what makes some children resilient, despite experiencing severe stressors? A child's genetic vulnerability in combination with environmental factors seems to affect the risk–resilience balance. So as with many other things, it is likely that a complex mix of gene(s)–environment(s) interactions is likely to underlie addiction vulnerability and development in our children. While this is not a fine science, much effort is now going into understanding the mechanisms involved in order to help us identify which children may be at greater risk of alcoholism, thereby enabling us as parents and society to reduce that risk.

## Lightweight v. Heavyweight

Teenagers often ask me why some of their peers ('heavyweights') can handle their drink more than others ('lightweights'). They ask this less out of curiosity than a desire to become heavyweights too. And they're prepared to go into training to achieve this.

It has now emerged that this aspect of alcohol tolerance is a strong predictor of later heavy drinking and alcohol-use disorders (AUDs). A study in 2009 by the Department of Psychiatry, University of California, which followed teenage and young men for twenty-five years, has found strong evidence that a person who has a low level of response (LR) to alcohol, meaning relatively little reaction to it, has a higher risk for developing AUDs. The researchers believe: 'If a person needs more alcohol to get a certain effect, that person tends to drink more each time they imbibe. Other studies we have published have shown that these individuals also choose heavy-drinking peers, which helps them believe that what they drink and what they expect to happen in a drinking evening are "normal" ... This low LR, which is perhaps a low sensitivity to alcohol, is genetically influenced.' They found that even when other predictors of developing alcohol problems were controlled for, LR was an independent, unique, consistent and powerful risk factor in predicting alcohol problems:

> Because alcoholism is genetically influenced, and because a low LR is one of the factors that adds to the risk of developing alcoholism, if you're an alcoholic, you need to tell your kids they are at a four-fold increased risk for alcoholism. If your kid does drink, find out if they can 'drink others under the table', and warn them that that is a major indication they have the risk themselves. Keep in mind, however, that the absence of a low LR doesn't guarantee they won't develop alcoholism, as there are other risk factors as well. We are looking for ways to identify this risk early in life, and to find

ways to decrease the risk even if you carry a low LR ... so there is hope for the future.[39]

And predictably, geneticists are busy doing precisely this, producing news headlines such as, '"Tipsy" alcohol gene could curb alcoholism'. Researchers believe they have now identified a gene (CYP2E1, known for its involvement with alcohol metabolism) that helps to explain why some people feel alcohol's effects more quickly than others.[40] The US researchers believe that 10–20 per cent of people have a version of the gene that may offer some protection against alcoholism. That is because people who react strongly to alcohol are less likely to become addicted. There is currently speculation that ultimately, people could be given CYP2E1-like drugs to make them more sensitive to alcohol – not to get them drunk more quickly, but to put them off drinking to inebriation. In practice, this means you will get more of a kick from champagne for half the price.

Another study has now identified a particular *network* of genes (glutamate receptor-signalling genes) as being strongly implicated in determining our LR to alcohol.[41] And a further study has found networks of genes that influence our degree of alcohol *consumption* that are different from the networks which predispose us to alcohol dependence: 'The genetic factors that contribute to the full range of alcohol consumption versus alcohol dependence in humans are distinct.' They believe that people with a set of genes that predisposes them to drink moderate amounts of alcohol may still have the genetic predisposition to lose control over their drinking behaviour, and perhaps become alcohol *dependent*. While people with a genetic predisposition to drink high amounts of alcohol may not have the genes that predispose them to become dependent.[42]

And so it appears that when it comes to handling your drink, teens will have to accept that you're basically born a heavyweight.

## Risk Reduction

We've looked at key aspects of parenting here that may increase the likelihood that children will develop a problem with alcohol. Of course it's easy to point to people who were given alcohol as children or allowed to drink as teenagers and who have grown up perfectly well, without any alcohol issues. (It's also true, by the way, that most infants who breathe in daily bedroom passive cigarette smoke and most children who smoke a packet of cigarettes every day will never develop lung cancer, that most children who snort cocaine will never become coke addicts, that most children who have unprotected sex will never acquire HIV and that most children who eat lumps of pig lard daily will not die prematurely from coronary heart disease.) But health policies and advice are based on informing us as parents of behaviours that increase or reduce the *likelihood* that something bad will happen to our children. Once aware of general risks and probabilities, we can make *informed* decisions and choices about our children, as opposed to succumbing to fashion-led acquiescence.

The studies cited in this chapter highlight a need and justification for us as parents not to pander to what our children are interested in, but to do what we instinctively feel is in their best interests. Whether our children agree with it or nor not, they do absorb our views and values about a wide variety of things, including drinking. *What we believe and expect as parents and as a society has a significant influence on how early, how often and how much our children drink.*

So returning to 'the French problem': buy their wine, not their weaning.

## Four

# Firewater on the Brain

In protecting our children from a drinking culture, we have to reconsider how we define the terms child and adult, because when it comes to alcohol the definitions differ widely within everyday or legal use. And there seems an inherent inconsistency: while a child legally becomes adult at the age of eighteen, the regions of their brain important for judgment, critical thinking and memory do not fully mature until they are in their mid-twenties. At the same time, a new generation of research is finding that alcohol can damage the normal growth and development of a teenager's brain cells in these regions.

So we have a problem on our hands.

As teenagers mature into their final year of secondary school, many parents begin to feel more comfortable about letting them drink alcohol. However, it now seems that loosening the reins on drinking may not be a good idea in the long run. In the US, the

Surgeon General's 'Call to Action to Prevent and Reduce Underage Drinking' in 2007 made this explicitly clear: 'Underage [under the age of twenty-one] drinking can cause alterations in the structure and function of the developing brain, which continues to mature into the mid- to late twenties, and may have consequences reaching far beyond adolescence.'[1]

While 80 per cent of brain growth takes place between birth and three years, the size, shape, complexity and function continue to develop in highly important ways for many years to come. A child's brain is 'plastic' in that it is constantly being physically shaped in response to their environmental experiences. And, like a clay sculpture, it ultimately 'sets'. The technical term for this process is structural neuroplasticity. One of the influences affecting the size, shape, complexity and function of a child and young adult's brain is alcohol.[2] And, as parents and a society, we want to prevent any distortion in the shape that our children's brains ultimately arrive at.

Understanding this point isn't merely an exercise in neurophysiology for the sake of it. These alcohol-related changes are linked directly to our children's intellect, personality, mental and physical health. Alcohol, even in small amounts, may have long-lasting effects on our children's brains that we simply didn't know about before.

Despite the hype from red-wine vintners, claiming that moderate drinking of their product will lead to a longer, healthier life, alcohol is actually neurotoxic, meaning it's poisonous to brain cells. This is certainly not a new discovery; even at the turn of the millennium the US Department of

Health and Human Services reported to the US Congress, saying: 'Studies clearly indicate that alcohol is neurotoxic, with direct effects on nerve cells.'[3] And it begins doing its work within only six minutes, according to a study in the *Journal of Cerebral Flow and Metabolism*, which describes in explicit terms exactly how alcohol may cause a harmful imbalance in the brain in favour of pro-oxidants, as opposed to antioxidants, making brain cells more vulnerable to damage by free radicals. The study describes 'binge drinking' in rather unflattering terms, as inducing 'immediate and toxic effects'; involving 'distinct biochemical and neurotransmitter changes' in the brain.[4]

Neurochemists – a profession not known for its outbursts – are increasingly vocal in their concerns about our society's misunderstanding of the actual effects of binge drinking: 'Urgent action is therefore needed to comprehend the aetiology and pathogenesis of the binge-drinking culture, as well as to educate individuals on the dangers of such drinking. There is little doubt that the problem of binge drinking, particularly by adolescents, needs to be addressed urgently, to prevent cognitive impairment, which could lead to irreversible brain damage.'[5]

In teenagers who drink regularly, the parts of the brain important in emotional and impulse control – the prefrontal cortex – have actually been found to be smaller and to remain so even at the age of twenty-one. And in teenagers who only binge drink (say four or five drinks) once a month, brain cells in many different parts of the brain are now found to be subnormal.[6]

## White Matters

'White matter' refers to brain areas that appear light in colour due to the brain cells being heavily coated in white protective fats. These cells are fibres that connect different parts of our brains. The health and condition of this white matter is essential to the efficient relay of information within the brain and is linked to performance on a range of cognitive tests, including measures of reading. Abnormalities in the health of the brain's white matter could affect our ability to think, to process information, function emotionally and make decisions. A study entitled 'Altered White Matter Integrity in Adolescent Binge Drinkers' has found that even in teenagers who have no history of any alcohol use or mental disorder and only binge drink infrequently (e.g. four or five drinks once a month), brain cells in eighteen parts of the brain are found to be thinner, weaker with less protective coating, leading to poor, inefficient communication between brain cells.' And the good news continues: '... infrequent exposure to large doses of alcohol during youth may compromise white matter fibre coherence'. One of the researchers commented, 'These results were actually surprising to me because the binge-drinking kids hadn't, in fact, engaged in a great deal of binge drinking. They were drinking on average once or twice a month.'[7]

And there seems to be direct intellectual consequences to these changes in the integrity of teenagers' white matter. In 2009, researchers looked at twelve- to fourteen-year-olds before they used any alcohol or drugs. Over time, some of the adolescents started to drink, in some cases four or five drinks per occasion, two or three times a month — classic binge drinking behaviour in teenagers.

Comparing the young people who drank heavily with those who remained non-drinkers, researchers found that the binge drinkers did worse on thinking and memory tests. There was also a distinct gender difference. The girls who drank did less well on tests of spatial functioning, which affects things such as mathematics and engineering functions, while the boys displayed poor performance on tests of attention – especially sustained attention (concentration). The lead researcher spelled out the practical implications for parents of teenage pupils who were hoping to get their children into good universities. 'The magnitude of the difference is 10 per cent. I like to think of it as the difference between an A and a B.'[8]

The study concluded that this type of moderate binge drinking during adolescence 'may adversely influence neurocognitive functioning. Neurocognitive deficits linked to heavy drinking during this critical developmental period may lead to direct and indirect changes in neuromaturational course, with effects that would extend into adulthood.'[9] And so, while teenagers vary in the way alcohol may affect their brains, for some teenagers there may be no 'safe' level of alcohol use. The investigators found negative effects in thinking and memory in teenagers after as little as twelve drinks a month, or two or three binge-drinking episodes a month.[10]

## Grey Matters

The parts of the brain dominated by nerve-cell bodies not covered in the white protective fat which look greyish in colour are referred to as grey matter. And alcohol seems to operate on an equal-opportunities basis in this respect – colour blind when it comes to brain matter.

A team at the University of California, San Diego, School of Medicine examined the hippocampus in fifteen- to seventeen-year-olds. (The hippocampus is a crucial area for memory formation and learning and it actively develops during adolescence.) All the teens came from middle- to upper-class families. One group misused alcohol in that they 'were primarily weekend binge drinkers' (with about seventeen days in between drinking episodes) and it was this group that 'had significantly smaller left hippocampal volumes' – in other words this key part of their brain was noticeably smaller than in the non-binge-drinking teenagers.[11] A previous study of fifteen- to eighteen-year-olds made a related finding of smaller left hippocampal volumes in the binge drinkers,[12] while another looked at a wider age range (thirteen to twenty-one years) and was equally cheerful: ' ... these findings suggest that, during adolescence, the hippocampus may be particularly susceptible to the adverse effects of alcohol ... toxic mechanism is age dependent and becomes fully sensitive between puberty and adulthood'.[13]

These differences in brain size are now associated with

abnormal brain functioning in teenage binge drinkers who haven't had a drink in over a month. Reflecting their abnormal brain scans, the teenage drinkers did less well on learning verbal material than their non-drinking counterparts.[14]

In trying to understand precisely what happens in the hippocampus when teenagers drink alcohol, another team of scientists reported in the *Proceedings of the National Academy of Sciences* in 2010 that in addition to being emotionally touchy, 'adolescence is a period of high vulnerability to brain insults'. They found that the hippocampus in adolescent macaque monkeys who binge drink became damaged in two ways. Alcohol seemed to interfere with the natural division and migration of the hippocampal brain cells. Furthermore this lasting alcohol-induced reduction in brain-cell production and development was accompanied by an increase in brain-cell degeneration. And they found a lasting effect, two months after the monkeys' last drink, which they believe may underlie the deficits in cognitive tasks, including verbal learning and memory.[15]

Other scientists are even more specific in pinpointing the precise chemical chain of events that damages brain cells in the hippocampus during binge drinking. While investigating the 'neuropathological' and immune system changes induced in the brain they found 'neuro-inflammation induced in the hippocampus of "binge-drinking" rats'.[16]

What remains unknown is if the cognitive downward slide in teenage binge drinkers is reversible.

## Size Matters

In 2008, a study of 1839 adults published in the *Archives of Neurology* found that the more alcohol an individual drinks, the smaller his or her total brain volume. Although 'most participants reported low alcohol consumption ... There was a significant negative linear relationship between alcohol consumption and total cerebral brain volume.' Although men were more likely to drink alcohol, the association between drinking and brain volume was stronger in women. This could be due to biological factors, including women's smaller size and greater susceptibility to alcohol's effects. The neurologists and epidemiologists wrote: 'The public health effect of this study gives a clear message about the possible dangers of drinking alcohol.'[17] We should ask ourselves why while newspapers are full of claims that alcohol will extend your life, most people will probably never have even heard of this large study.

In fifteen- to seventeen-year-olds who regularly drink alcohol, parts of the brain important in emotional and impulse control have actually been found to be smaller. 'Consistent with adult literature, alcohol use during adolescence is associated with prefrontal volume abnormalities, including white matter differences.' In particular, females with alcohol-use disorders have been found to have a smaller prefrontal cortex (PFC).[18] A previous study looked at a wider age range of thirteen- to twenty-one-year-olds and found those 'with alcohol-use disorders had smaller prefrontal cortex and prefrontal cortex white matter volume'.[19]

Until we fully understand the relationship between alcohol consumption and a smaller prefrontal cortex in children and young people, there are exceedingly good reasons why we must err on the side of caution. The ability to delay gratification allows humans to make decisions and accomplish goals. This vital function is rooted in a part of the frontal lobe of the brain: the prefrontal cortex.[20]

In particular, the lateral prefrontal cortex is critical for making decisions in which forgoing a small immediate reward can lead to a better future outcome. On the other hand, an inability to delay gratification is implicated in psychiatric disorders related to impulse control such as substance abuse. The pattern of impairments in people with antisocial personality disorder with highly psychopathic tendencies shows a remarkable resemblance to that in people with frontal-lobe damage, suggesting that an underlying cause of the behavioural disturbances seen in psychopathy may be dysfunction in the frontal lobes.[21]

In children and adolescents, the lateral prefrontal cortex is not yet fully developed. This helps to explain why these younger age groups have a harder time delaying gratification.

Higher intelligence is related to better self-control, and the anterior prefrontal cortex, which is one of the last brain structures to fully mature, is heavily involved. Better self-control is relevant to a host of important behaviours, ranging from saving for retirement to maintaining your physical and mental health.[22]

These findings provide a glimpse of the neurological landscape through which alcohol travels, preventing teenagers from developing self-control when they're sober because their brain hardware is underdeveloped. Whether in moderation or excess, adults drink alcohol because it disinhibits those parts of the brain that have evolved over many years to control thoughts, feelings and behaviours. Of course, most of us prefer to view this less technically in terms of 'relaxing' and 'enjoying ourselves'. And, in most cases, this is what it does. Yet alcohol's extraordinary ability to disinhibit our other adult impulses is well documented every day of the week. And we need only look at the role of alcohol in the antisocial behaviour of adults to realise that there's no reason why it should not help our children to become badly behaved too. But the effects on the young are, in fact, far more fundamental: when children and teenagers drink, the disinhibition that takes place does so precisely at the point when their brains and behaviours are still undergoing crucial development of the ability to control impulses, so disrupting this essential process. The temporary disinhibition that results from a few hours of drinking may lead to a longer-term general disinhibition while children are sober, ultimately becoming a permanent feature of their characters.

A study published at the end of 2010, has found that teenagers who binge drink are more likely to engage in risk-taking behaviour later on. It is already known that impulsive behaviour can be caused or exacerbated by drinking alcohol within a few hours time period, but new findings suggest that it may increase the level of risk taking and impulsive behaviour over a long-term period. The researchers believe that because adolescence is a time when many begin to drink, this can have serious effects on brain development. In studying male teenagers and young adults they identified a significant trend regarding the amount of alcohol an individual drinks, and changes in levels of impulsive behaviour that follow the *next* year.

The study, by the Center of Alcohol Studies at Rutgers University, involved annually following more than five hundred boys aged eight to eighteen, with another follow-up at ages twenty-four and twenty-five. What is particularly striking is that it is the vast majority of teenagers and young people who are only *moderately* impulsive that are most affected in the long term. The results showed that for adolescent boys exhibiting moderate levels of impulsive behaviour (61 per cent of all of the young adults), as opposed to those in the low or high groups, there was a significant increase in impulsive behaviour when they'd engaged in heavy drinking the previous year: 'heavy drinking may increase impulsive behaviour by affecting the development of brain areas that support behavioural control or through other associated mechanisms'.[23] 'Heavy alcohol use in adolescence may lead to alterations in brain structure and function that reduce behavioral (impulse) control, which could, in turn, promote further heavy drinking ... These studies highlight the importance of prevention. Decreasing heavy drinking during adolescence may decrease impulsivity by preventing damage to crucial brain areas.'

Yet adolescent and teenage parenting guides today tend to treat underage drinking as a *consequence* or a symptom of an underlying problem, rather than something that can in its own right actually change neurologically – the way a child behaves and develops in future. And so, by allowing our children to learn to let go and 'chill out' with alcohol, we may, paradoxically, be making rods for our own backs.

Even the largest white matter in the brain, the corpus callosum – the connecting cable for communication between the left and right halves of the brain – seems to be affected by alcohol. In seventeen-year-olds with alcohol-use disorders (AUDs), the integrity of the corpus callosum was affected 'suggesting neurotoxic effects of alcohol on adolescent corpus callosum microstructure'. Girls were worse off they said: 'As seen in adults, female adolescents with AUD may be especially vulnerable to corpus callosum mircostructural injury.'[24]

Binge drinking has been found to cause changes in gene expression within the brains of animals. The Institute of Psychiatric Research, Indiana University, looked at genes in the amygdala, a part of the brain which plays a key role in the processing of emotions and is linked to both fear responses and pleasure. Conditions such as anxiety, autism, depression, post-traumatic stress disorder and phobias are all suspected of being linked to abnormal functioning of the amygdala, owing to damage, developmental problems or neurotransmitter imbalance. It may sound complicated, and there's no need to understand the fine detail, but the conclusion of this study should make it clear that alcohol continues to have new-found consequences of which parents and politicians are completely unaware: 'Overall, the results indicate that binge-like alcohol drinking ... produces region-dependent changes in the expression of genes that could alter transcription, synaptic function and neuronal plasticity.'[25]

## Brain Function

While the effects outlined above focus on the size, structure, gene function and chemistry of the brain in relation to alcohol, scientists have also been looking at electrophysiological changes, better known as 'brain waves'. In 2010, the Molecular and Integrative Neurosciences Department of the Scripps Research Institute reviewed what is known about the electrophysiological effects of binge drinking in adolescent rats. They concluded that even relatively brief exposure to high levels of ethanol during adolescence in the rat 'is sufficient to cause long-lasting changes in functional brain activity'. Most disturbing was that brief exposure during adolescence was also found to produce 'long-lasting reductions in the mean duration of slow-wave sleep (SWS) episodes and the total amount of time spent in SWS, a finding consistent with a premature ageing of sleep'.[26]

Slow-wave sleep is the deepest stage of our nightly sleep and the most 'restorative'. For example, when scientists selectively suppressed SWS in young, healthy, lean people, even though they slept for the same period of time as normal, the scientists reported in the *Proceedings of the National Academy of Sciences* that they found 'marked decreases in insulin sensitivity ... leading to reduced glucose tolerance and increased diabetes risk'. They wrote that the magnitude of the change in insulin sensitivity was comparable with that associated with a weight gain of two stone![27]

## Why They Drink

Nature made the brains of children excitable. Their neurochemistry is ready to respond to everything in their environment. On the positive side, this is partly why children can learn so easily. The adolescent brain too is primed and ready for intense, all-consuming learning; this is a unique feature that drives much adolescent behaviour. Young people become very passionate about a particular activity or sport, about music or 'changing the world'. This is, however, a two-edged sword because those same inclinations to explore and try new things may also increase the likelihood of doing risky or negative things: driving fast, sex, sex, sex, nicotine, cannabis, cocaine, ecstasy and, of course, binge drinking.

Alcohol dependence is a form of neurochemical 'learning'. An adolescent's brain is designed to form new connections in response to the environment; so if potent substances suddenly enter that environment, they are aiding and abetting a much more powerful habit-forming inclination and ability than that found in adults. (See Chapter 8 for a more detailed discussion of the role of risk seeking.)

Research in developmental psychobiology is finding that adolescents are more sensitive than their adult counterparts to the positive rewarding effects of alcohol and drugs, while unfortunately they're less sensitive to the aversive properties of such stimuli. And this gift may be exacerbated further by a history of prior stress or alcohol exposure, as well as by a genetic disposition which may then enable a relatively high level of adolescent alcohol use and 'an increased probability for the emergence of abuse disorders'.[28]

## Getting from A to B

A major study by the US Department of Health and Human Services found that approximately two-thirds of high-school students with 'mostly As' are non-drinkers, while nearly half of students with 'mostly Ds and Fs' report occasional binge drinking;[29] while a study by the British government in 2010 found that 'drinking alcohol was associated with lower GCSE scores and being not in education, employment or training (NEET)'.[30] A decade earlier, the Department of Physiology, University of Granada, studied alcohol consumption and academic performance in a population of Spanish secondary-school students aged fourteen to nineteen and concluded: 'Although we cannot draw any conclusions about the causes of the association between academic failure and teenage drinking, our results do show that the risk of failing increases together with alcohol intake.' They also found that there's no such thing as a better class of drunk, the link between alcohol and academic failure applying equally to students at both state and private schools.[31]

In the United States, the Division of Epidemiology and Prevention Research of the National Institute on Alcohol Abuse and Alcoholism looked at the grade average of students aged seventeen or older and found that 'relative to non-drinkers, students who started drinking before age thirteen were 46 per cent more likely to obtain at least one letter grade lower' and even in the present, those who had been binge drinking in the preceding thirty days were between 17 and 32 per cent more likely to have lower grades. Their conclusion is good, solid advice: 'The strong association between alcohol use and school performance found among US high-school students underscores the need to delay drinking onset and binge drinking in this age group.'[32]

Some social psychologists and therapists may argue that underage drinking appears to be largely a symptom, rather than a cause, of poor academic adjustment – that students already doing well in school are less likely to drink, whereas those doing poorly are more likely to do so.[33] But this does not discount the growing findings on brain changes and cognitive deficits strongly linked to drinking in adolescents and young people. Both can be true.

At university, the negative relationship between alcohol consumption and grades continues. A study of 9931 students carried out by the University of Minnesota Health Service found a linear relationship – as binge drinking goes up, grade-point average (GPA) goes down (from an average GPA of about 3.30 for students who had not engaged in binge drinking in the preceding two weeks to about 3.10 to 3.15 for those who had done so twice or more in the previous two weeks).[34] However, at this age the relationship between drinking and grades may be more complicated. Many students live away from home, and living environments (i.e. other students in the halls of residence) could play a big role in increased drinking and, therefore, have an impact on grades. On the other hand, drinking among some students may affect the living environment and thereby affect other students' grades.[35]

In almost all other areas of child health, our society errs on the side of caution, subscribing to the ancient medical principle, 'First, do no harm'. Given the vulnerability of developing brains, the overriding importance of proper brain development, the powerful implications of distortions or hindrances to that development and the growing link between alcohol and significant changes for the worse in the brains of children and young people, we must clearly reduce and prevent exposure to alcohol while brains are at this formative stage. To continue to keep such an open mind on the subject is a clear sign that one's brains have fallen out.

## Five

# Girls Just Wanna Have Fun?

This chapter will reveal an unexpected form of sexual discrimination ordained by Mother Nature – and one that no tribunal can remedy.

Real-life experience and unnecessary research studies confirm that when it comes to accepting an offer of a one-night stand for sex from a woman, men are very open-minded. In fact, the majority of men have almost non-existent standards. Unlike women, men operate more of an equal-opportunity policy when it comes to sexual partners. In 2009, a study conducted in three countries of standards required for a one-night sexual encounter found that men were more than eighteen times more likely than women to accept a bedroom offer from an unattractive person. 'Men and women demonstrated a striking difference in interest in casual sex.' This is explained by the researchers in terms of 'differences in mating strategies'.[1] In one classic study, people were approached by members of the opposite sex and propositioned. 'The great majority of men were willing to have a sexual liaison with the women who approached them. Women were not. Not one woman agreed to a sexual liaison.'[2]

It has been generally accepted in all parts of the world I've visited that being a young man with a libido is like being handcuffed to a maniac. And the experiments above merely add the numbers to verify what we knew was true all along.

## Drink 'Til She's Cute

When it comes to casual sex, women do the opposite of men, raising their standards with respect to a potential short-term mate's attractiveness. However, alcohol changes things.

There's a T-shirt displaying the sage advice for young men, 'Drink 'til she's cute, but stop before the wedding'. (However, my advice to boys would be: 'If she looks like the back end of a bus when you have your first drink, she'd better still look like the back end of a bus when you've had your last or you've had far too much to drink.') And a female counterpart version of the T-shirt is now on sale, its message, printed in full capital letters, shouting, 'DRINK 'TIL HE'S CUTE!' (available in pink, lime and yellow), along with an alternative version: 'DRINK UNTIL HE'S CUTE' (available in light blue) modelled by a hairy man with a full beard and moustache wearing a woolly hat.

Scientists are now actually studying the mechanisms behind the T-shirt, publishing serious academic papers with titles such as, 'An explanation for enhanced perceptions of attractiveness after alcohol consumption', containing findings such as:

> Acute alcohol consumption increases ratings of attractiveness to faces. This may help to explain increased frequencies of sexual encounters during periods of alcohol intoxication. At least in part, such increased attraction may be the result of alcohol consumption decreasing ability to detect bilateral asymmetry, presumably because of the reductions in the levels of visual function ... The reduced ability of inebriated people to perceive asymmetry may be an important mechanism underlying the higher ratings of facial attractiveness they give for members of the opposite sex and hence their increased frequency of mate choice.

In other words they drank 'til she/he was cute, or at least not so ugly.[3]

Those readers who spend a significant proportion of their money on make-up may feel either horrified or suddenly rich after reading this: a 2009 study published in the *British Journal of Psychology* showed that while men found adult women (who were wearing make-up) more attractive after consuming alcohol, the alcohol did not interfere with their ability to determine a woman's age. Or, as the empiricists put it, 'Alcohol consumption does not interfere with age-perception tasks in men.' Apparently, it's because you're not worth it.[4]

## The Full Monty

Alcohol education and debate has generally been a separate affair from the more emotive debates about casual sex, HIV transmission, underage pregnancy, rape and date rape. However, since 2005, the health establishment has begun to recognise the strong connections, and they're extremely concerned.

In 2005, the World Health Organization published 'Alcohol Use and Sexual Risk Behaviour: A Cross-Cultural Study in Eight Countries',[5] in which it states: ' . . . alcohol is commonly used as a disinhibitor, a sex facilitator . . . The synergy between sexual behaviour and alcohol use enormously multiplies the potential negative consequences of the two behaviours separately . . . Alcohol use has also been linked to early sexual experiences . . . the use of alcohol at/during (first) sexual encounters.'

Unfortunately, women around the world have a lot in common: 'Among women, alcohol use increases involvement in risky sexual encounters and sexual victimisation, exposing them to the risk of unwanted pregnancies and STIs . . . various interactive links between alcohol use and sexual behaviour – alcohol consumption not only presented as a "precursor" of risky sex, but also as an outcome of it.'

The WHO also noted that the taboos and stigmas for engaging in easy sex have been significantly eroded. 'The effect of modernisation and the media on the youth, which manifests in early drinking, early sexual activity and increasing vulnerability to risk behaviours ... The media (electronic and print) play an important role in shaping and influencing sexual behaviour and alcohol-use patterns.'

Alcohol, the grand sexual facilitator for adolescents, is now increasingly found to play a key international role. A cross-cultural study in nine European cities on youth and young adults reported that an epidemic of binge drinking 'exposes millions of young Europeans to routine consumption of substances which alter their sexual decisions and increase their chances of unsafe and regretted sex'. Much of the drinking, according to the study, is strategic – for example 28.6 per cent of alcohol users use it to facilitate sexual encounters. And there was also a strong link between alcohol and early sex, particularly among girls.[6]

Another study in Britain, published in the delightfully titled monthly *International Journal of STD & AIDS* in 2007, rightfully complains: 'The ancient Greeks had a single god, Dionysius, for sex, wine and intoxication, and associations between alcohol intoxication and sexual risk would appear to be self-evident. However, alcohol does not feature in the UK Strategy for Sexual Health and HIV.'

The study reported 'particularly marked' increases in young women going to STD clinics, attributing this to the big increase in drinking. 'The UK has a specific problem with binge drinking that is not seen elsewhere in Europe, and it seems highly likely that this binge-drinking culture is a risk factor for all forms of sexual risk, including sexually transmitted infections (STIs).'

The researchers monitored the drinking patterns of patients at a typical STI clinic in a large city in the south of England and found 'the majority are binge drinking to a significant extent ... In women, there was a significant correlation between the number of sexual partners and both frequency of drinking days, weekly intake and binge drinking. There was no correlation with binge drinking in men and a less marked correlation with overall weekly intake.' They went on to report that 76 per cent of women had experienced unprotected sex as a result of drinking, and women who binged most heavily experienced significantly more unwanted pregnancies. A fifth of the women, 'reported an unwanted pregnancy, with 28 per cent drinking beforehand'.[7]

And it's a familiar picture on the other side of the world too. In 2010, the Preventive and Social Medicine Department of the University of Otago, studied 2548 undergraduate students at five New Zealand universities and found that those who drank the most heavily were ten times more likely to have unsafe, or unprotected, sex. Students who began drinking heavily while at secondary school were likely to continue at university and participate in risky sex experiences, including unsafe sex or sex they regretted. Half of the university students had had their first alcoholic drink when under the age of fifteen. Those who started later, drank less at university and were less likely to have unsafe or unwanted sexual experiences.[8] In Australia, an eleven-year study in the *Australian and New Zealand Journal of Public Health* reported, 'Rates of alcohol consumption among secondary students aged fifteen to seventeen have increased markedly, as has the proportion of young people engaging in sex while under the influence of alcohol ... and increased rates of sexually transmitted infections.'[9] And back in New Zealand again, a thirty-year prospective longitudinal study found 'a straightforward existence of a causal pathway' in which the more you drink, the greater your direct risk of contracting an STD.[10]

Even as I write, a new study of 25,000 people carried out over a ten-year period is painting a similar picture. Researchers from University College London and the Health Protection Agency report 'alcohol use as an important determinant of early sexual activity'. Women who drank more than 'the recommended' fourteen units a week were 80 per cent more likely to have taken emergency contraception such as the morning-after pill at least once over the last year. They were also 40 per cent likelier to have had at least one abortion in the last five years.[11]

At the same time, official statistics show that the number of abortions in the UK has soared since the 1970s, making Britain the termination capital of Europe. And our children are the most likely to get drunk.

I mentioned this issue to a nurse at one school last week and a house mistress at another this week. Both told me, independently, that increasingly they have younger and younger girls approach them on a Monday morning asking for the morning-after pill. But when the girls are asked whether they had actually had full sexual intercourse, they can't even remember where they had been, whether or not they had even kissed any boys or, indeed, anything about the evening in question. They simply want the morning-after pill in case they had a sexual encounter(s) that they can't remember. One girl could apparently only say, 'When I woke up yesterday, my tights were ripped, Miss.'

The UK's teenage pregnancy rate is almost the highest in the Western world, second only to the USA's. Those who fuss about these statistics should also be looking at the co-conspirator – alcohol – as the object of their wrath and concern.

## The Legless Leg-over

The erosion, and in countries such as Britain, the exodus of stigma and social embarrassment of young females having casual sex, or even being a single parent, has conspired with higher and earlier consumption of alcohol to increase the statistics.

One of the contemporary indulgent behaviours decried by older generations is casual sex – even sex on the open pavement – among younger and younger girls, along with all that goes with it. Parents and schools are desperate to regain some authority over children's sexual behaviour. And while much debate surrounds how much sex education our children should have and at what age, few people understand the enormous role that alcohol plays in sexual behaviour. Which is extraordinary, considering that there are few male parents, teachers and politicians who haven't used alcohol strategically to get their leg over.

## Spiked Drinks?

The latest misguided witch hunt focuses on 'spiked' drinks and 'date-rape' drugs. Yet consultants and biochemists who have conducted medical studies of blood and urine samples from young females attending hospital accident and emergency departments, claiming that their drinks had been 'spiked' with date-rape drugs, are concluding: 'No one tested positive for rohypnol or GHB. The symptoms are more likely to be a result of excess alcohol.'[12]

Similarly, an Australian study at two hospitals in Perth found that none of the ninety-seven young men and women who claimed to have had their drinks spiked had been drugged. 'Our findings do not support the public perception that sedative drugs are being used to spike people's drinks. We did not identify a single case where a sedative drug had been placed in a drink in a pub or nightclub setting.' Yet, 'Thirty-five per cent of patients still believed that they had been a victim of drink spiking irrespective of the results.'[13]

And criminologists are coming to the same conclusion. A study in 2009 in the *British Journal of Criminology* on students in the UK and US stated: 'There is a stark contrast between heightened perceptions of risk associated with drug-facilitated sexual assault (DFSA) and a lack of evidence.' And, they noted, 'There is widespread disbelief, or active denial, that excess alcohol could cause the same incoherence, physical distress and incapacity associated with "date-rape" drugs.' The researchers commented that 'young women appear to be displacing their anxieties about the consequences of consuming what is in the bottle on to rumours of what could be put there by someone else'.

The researchers are interested in knowing whether the 'urban myth' of drink spiking is also the result of parents feeling unable to discuss with their young adult daughters how to manage drinking and sex and expressing their anxieties about this through talk of drink-spiking risks.[14] In the meantime, the drinks industry is exceedingly unlikely to be upset by the perpetuation of this 'urban myth'.

So it seems we find it hard to accept the obvious: that drinks do not have to be spiked with 'date-rape' drugs for women (and even men) to be drugged and date-raped. Alcohol has always been a highly effective date-rape drug in its own right.

Obviously, the other part of this equation is that boys' libido, along with their impulse control, is also powerfully disinhibited when they're drunk, years before that impulse control is fully formed (see p. 92). They are more likely to 'date-rape', or be falsely accused of either that or sexual assault, if they have been drinking. As covered in further detail later on, legally a woman may not be capable of giving consent to sex if she is drunk, even if she is conscious and seemingly 'gagging for it'. The previous government made it clear again that they want the conviction rate for rape to increase considerably. This will be discussed in Chapter 8.

## Blackouts

The misplaced belief in the spiked drink and date-rape drug phenomenon begs the question of exactly *why* women can't fully remember what happened the night before.

Alcohol can produce noticeable impairments in memory after only a few drinks and, as the amount of alcohol increases, so does the degree of impairment. Memory 'blackouts' are much more common among social drinkers than previously assumed. In 2002, the Department of Psychiatry at Duke University surveyed 772 undergraduates about their experiences with blackouts and asked, 'Have you ever awoken after a night of drinking not able to remember things that you did or places that you went?' Of the students who had ever consumed alcohol, 51 per cent reported blacking out at some point in their lives, and 40 per cent reported experiencing a blackout in the year before the survey. Of those who reported drinking in the two weeks prior to the survey, 9.4 per cent said they had blacked out during that time. The students admitted to learning subsequently that they had participated in a wide range of potentially dangerous events they could not remember, including vandalism, unprotected sex and dangerous driving.

Despite the fact that the men drank significantly more often and more heavily than the women, equal numbers of men and women reported experiencing blackouts.[15] This shows that regardless of the amount of alcohol consumption, females – a group rarely studied in the literature on blackouts – are at greater risk than males for experiencing them. The tendency for a woman to black out more easily than a man probably results from differences in how they metabolise alcohol.

Females also may be more susceptible than males to milder forms of alcohol-induced memory impairments, even when men and women drink comparable amounts of alcohol.[16] A study of

8833 fifteen- to sixteen-year-olds, carried out in England in 2009, says it all: 'Our results, even after correcting for binge and drinking frequency, identify an independent association between tendency to forget things after drinking and being female. Such damage may now be exacerbated by young females' consumption of alcohol in the UK approaching the same level as males.'[17]

The effects of alcohol on female memory are in keeping with good evidence that women may be particularly vulnerable to its effects on many key organ systems. For example, alcoholic women develop cirrhosis of the liver, alcohol-induced damage of the heart muscle (cardiomyopathy) and nerve damage (peripheral neuropathy) after fewer years of heavy drinking than do alcoholic men. And research comparing men and women's sensitivity to alcohol-induced brain damage find that alcoholic women have been drinking excessively for only about half as long as the alcoholic men studied.[18] This indicates that women's brains, like their other organs, are more vulnerable to alcohol-induced damage.

A study by the Stanford University School of Medicine in 2005 found a 'faster progression of the developmental events leading to dependence among female alcoholics and an earlier onset of adverse consequences'. They also found that brain atrophy seems to develop faster in women: 'The women developed equal brain-volume reductions as the men after a significantly shorter period of alcohol dependence than the men.'[19]

And to complete the picture, the Pavlov State Medical University, St Petersburg, found in 2007 that not only do women become alcohol dependent more quickly than men, alcohol impairs cognitive functioning in women (including perceptual and visual planning and processing, working memory and motor control) more severely than in men. 'Our findings confirm and extend prior research that alcohol exerts more profound adverse effects more quickly on women compared to men.'[20]

## Female 'Pharmacokinetics'

So why is there so much sexual inequality around addiction? There are several metabolic and physiological differences between men and women that are central to this issue, explaining how women's bodies treat (and are affected by) alcohol differently from men's.

Firstly, men have a higher proportion of body water (61 per cent) than women (52 per cent), allowing the absorbed alcohol to be better diluted in their body fluids. In addition, females have a higher proportion of fat, compared with males, and this increases with age. Alcohol isn't absorbed by fat, so it ends up in the non-fatty tissues of the body, including the brain, muscles and liver.[21] Plus, men are generally bigger (taller and heavier) than women, and – again, generally – the bigger a person is, the greater the dose of alcohol, drugs or medicine required in order to have an effect.

Secondly, males are able to deal with a greater metabolic load of alcohol than females because they generally have more of the key enzymes which break down alcohol (the alcohol-degrading liver enzymes, alcohol dehydrogenase and aldehyde dehydrogenase). These alcohol-degrading enzymes also appear to be more active in males than in females.

Although alcohol metabolism occurs mainly in the liver, it begins in the stomach, called first-pass metabolism. When more alcohol is metabolised in the stomach, less is absorbed directly from the stomach into the bloodstream and therefore less reaches the liver to be metabolised. Unfortunately, women also have fewer alcohol-degrading stomach enzymes than men and therefore less effective first-pass metabolism of alcohol.[22] A study entitled, 'Gender Differences in Pharmacokinetics of Alcohol' found that: 'The gender difference in alcohol levels is due mainly to a smaller gastric metabolism in females.'[23]

The body's ability to break down, or 'detoxify' alcohol works most effectively below moderate levels of drinking, e.g. two standard drinks for someone weighing 70kg. But if we drink more than this, or if absorption is fast, the quantity of alcohol reaching our liver can exceed the metabolic capacity of the available enzymes. This allows a greater proportion of alcohol to escape first-pass metabolism and reach the general circulation, resulting in a higher peak blood-alcohol concentration (BAC). Women end up with higher BACs than men after drinking equivalent amounts of alcohol, even when the doses are adjusted for differences in their body weight.

Lastly, in addition, the organs and tissues of women appear inherently more vulnerable to the toxic effects of alcohol. It has been suggested that this is related to gender differences in oestrogen concentration.[24]

What all this means in practice is that girls and grown women are far more easily affected both on the night and for years in the future by the alcohol they drink now. They get more drunk more rapidly than men. To counter this, women need always to drink at least a third less than men for their entire lives. And a new generation of girls has to be brainwashed with this home truth to protect them from the growing menu of harm outlined above.

What we need is for the EU to devise a Man Pint and a Woman Pint, along with sexually unequal wine glasses.

## Future Fertility

While worrying about alcohol leading to unwanted pregnancy, we have also to face uncomfortable evidence that alcohol can damage future fertility. Professors of cell biology, neurobiology and anatomy working for the US National Institutes of Health summarised some of alcohol's effects on female reproductive function thus: 'Alcohol use negatively affects puberty in females, disrupts normal menstrual cycling and reproductive function ... Rapid hormonal changes occurring during puberty make females especially vulnerable to the deleterious effects of alcohol exposure during this time. Thus, the high incidence of alcohol consumption among middle-school and high-school students in the United States is a matter of great concern.'[25]

Although there is a growing understanding of the way that even small amounts of alcohol can affect the growing foetus, there is good reason to be concerned about the eggs before the foetus is even created. Unlike boys, girls are born with their egg supply for life. By birth, the number of egg cells (oocytes) numbers 1–2 million, and then drops to 300,000 by puberty. Only a small percentage of oocytes mature into eggs, with only about 400 eggs being released during a woman's reproductive life.

The egg is one of the longest-lived cells in the body and it appears that it can be affected by drinking. For example, researchers have found that in the unfertilised eggs of mice exposed to the equivalent of one episode of 'binge drinking', exposure to alcohol could induce chromosome segregation errors in the ovulated oocyte. Those eggs fertilised had a high chance of being spontaneously aborted or of resulting in babies with moderate to severe degrees of mental retardation, cranio-facial and other abnormalities, as well as a significantly reduced life expectancy. 'The findings from our experimental studies that have been carried out in mice,' say the researchers, 'draw attention to important principles which are of general applicability to the situation in the human. The potential hazard of the exposure to alcohol of pre-ovulatory human eggs is at least as harmful as it is during pregnancy, and should, as such, be an equal cause for concern.'[26]

This specific topic was actually discussed in the House of Lords as long ago as 1987,[27] but the public are still in the dark.

## Pregnancy

At least half of all pregnancies in the UK (and in many other industrialised countries, including the US) are unplanned, and so many women who become pregnant remain unaware of the fact for several weeks or more. And if the mother is drinking alcohol, so is her unborn child. Drinking alcohol while pregnant causes foetal alcohol spectrum disorders (FASD), which the BMA Board of Science recently classified as ' ... the leading known cause of non-genetic intellectual disability in the Western world'. The prevalence rate is thought to be 0.5 to 2 cases per 1000 births. (As a point of comparison, Down's syndrome, which is far more recognised, has a prevalence rate of 1.1 per 1000 births.) And in mice, it's been found that blood-alcohol levels equivalent in a human to around one and a half times the legal driving limit in the UK and US chemically alter the foetus's DNA. Infant mice that were exposed to alcohol in the womb also had some of the symptoms of human foetal alcohol syndrome, such as a lower body weight and smaller skulls. This suggests that if women drink too much in pregnancy, epigenetic changes (that is, changes in the way in which important genes function) may cause some of the permanent symptoms seen in foetal alcohol syndrome in their children.[28]

In some cases, epigenetic changes in people seem to be passed on to subsequent generations.[29] And research about to be published concerning the brain of the unborn baby has identified 'structural changes in the caudate nucleus in offspring of mothers who drank moderate levels of alcohol during pregnancy'.[30]

The proportion of young female binge drinkers has doubled since 2002. In 2008, more than 5000 girls in England between the ages of ten and fifteen needed hospital treatment for alcohol poisoning . . . and they are the next generation of mothers.

We need to talk.

## Breast Cancer

A study in 2010 of almost 7000 girls, starting when they were aged between nine and fifteen years and lasting for eleven years, found that those girls and young women who drank alcohol increased their risk of benign (non-cancerous) breast disease. The researchers at Washington University School of Medicine in St Louis and Harvard University also pointed out that benign breast disease increases the risk for developing breast cancer. They said, 'The study is an indication that alcohol should be limited in adolescence and early adult years and further focuses our attention on these years as key to preventing breast cancer later in life.'

Many women begin drinking alcohol as adolescents – the very time at which their breast tissue is going through stages of rapid proliferation. And the researchers emphasise that 'Now it is clear that drinking habits throughout life affect breast cancer risk, as well.'[31]

## Bingeing Ain't Beautiful

Of all the risks that binge-drinking teenage girls and young women face, here is a final one to which many may pay closer attention.

A study of eighteen- to twenty-five-year olds ('What Men Want: The Role of Reflective Opposite-Sex Normative Preferences in Alcohol Use Among College Women') asked women how many drinks a typical male student would like his female friends to drink at an event. The researchers compared the women's estimates with reality and found that almost three quarters of the females overestimated the men's actual preferences; in fact, 42 per cent of their estimates were nearly double what the men preferred. The lead researcher offered this advice: 'Our research suggests women believe men find excessive drinking sexually attractive and appealing, but it appears this is a giant misperception.'[32]

When it comes to alcohol consumption, there is, clearly, no sexual equality.

Six

# Reaches the Parts Other Beers Cannot Reach

At one time, it was Americans who suffered not only from obesity but from the Euro-sneer as they waddled around European capitals in their polka-dot Bermuda shorts with Polaroid cameras hanging from their necks. Today, McDonald's is opening a McCafé at the Louvre, Paris, where you can now enjoy le Big Mac while looking at la McMona Lisa. Those of a more refined palate can dine at KFC Rue Wolfgang Amadeus Mozart or simply hit the hut at Pizza Hut Boulevard Saint-Michel.

So now in Europe, the fat lady does indeed sing, and no louder than in Britain whose fat lady is the pudgy prima donna of the EU. In their fat calculations for Adult Overweight and Obesity in the European Union,[1] the International Association for the Study of Obesity shows almost 57 per cent of English women as being overweight or obese while the Scottish lasses beat their English counterparts at almost 60 per cent.

But the fat man sings too. British men are the beer-bellied baritones of the EU – literally the fat man of Europe – and they're even fatter than their wives: 66 per cent of English and Scottish men are fat. When ranking countries according to big boys who have a body mass index (BMI) of 30 or over (obese), the Scots reign again, followed closely by the English – and they share the fattest beer belly in Europe.[2] The most recent report (2010) from the Organisation for Economic Co-operation and Development (OECD) finds, 'Obesity rates in the United Kingdom are the highest in Europe. In England, rates have increased faster than in most OECD countries ... and are projected by the OECD to rise a further 10 per cent during the next ten years.' This is one of the world's fastest-growing obesity rates, placing the UK in fifth place behind the US, Mexico, Chile and New Zealand.[3] British beer bellies increasingly rule ... not OK.

And in a study of citizens of the thirty-three richest countries, almost one in three children is now overweight in England and more than one in three in Scotland.[4] Levels of fatness in Scottish children are the second worst in the world, behind the US, while children in England are the eighth worst.

Children who have at least one obese parent are three to four times more likely to be obese themselves. This is partly genetic, but also because children generally share their parents' unhealthy diets and sedentary lifestyles. Yet alcohol rarely features in any discussion as to why adults are fat and how teenage drinking may help children to waddle in their parents' footsteps.

In looking at factors affecting illness rates and medical insurance, life insurance and critical-illness insurance, a study in 2007 by the life insurance company Standard Life found: 'The average adult in the UK is drinking an extra day's worth of calories every week through alcohol alone. The average adult is drinking enough lager, wine, cider and spirits to add almost 3000 calories to their weekly calorific intake. This is the equivalent to 500 calories above the average male recommended daily limit of 2500 calories and 50 per cent more than the advised maximum of 2000 calories a day for a woman.[5]

But it isn't simply the extra calories that are fattening up the nation. Alcohol changes other things, including our appetite. For centuries, wine has been coupled with food. The French word 'aperitif' refers to 'appetite' and describes an alcoholic drink believed to stimulate appetite. Scientists at the Mayo Clinic and Washington University[6] have begun to unpackage what is described as the 'wine paradox', a well-recognised phenomenon referring to the fact that people clearly do not eat less during a meal having drunk wine before it.[7] 'In other words, there is no compensatory reduction in caloric intake during a meal following alcohol ingestion. Individuals eat a full meal with "extra" calories coming from the alcohol.' Alcohol is classified as being orexigenic: it stimulates our appetite.

And scientists have stood by the dinner table to watch wine do its thing on our gluttony, capturing evidence of its orexigenic effects and exploring the possible mechanisms underlying our increased piggery. For example, in one often-cited study, scientists examined fifty-two people and evaluated several types of pre-meal drinks, including a dry white wine that contained 10 per cent alcohol (238 calories), as well as non-alcoholic drinks with the same number of calories that contained fats, proteins or carbohydrates. Also included in the study was an option for just plain water and/or no pre-meal drink. The investigators observed that food consumption was increased over a twenty-four-hour period after the dry white wine, as opposed to the non-alcoholic pre-meal drinks. With wine, the subjects actually ate a third more calories, their rate of eating was faster (44 grams of food per minute v. 38 grams per minute) and the duration of their meal was longer (14 minutes v. 12 minutes). In addition, it took almost two and a third times longer for them to reach satiation and eating after satiation continued for a period which was four times longer.[8]

And scientists continue to delve deeper to identify the hormones and neurotransmitters behind the appetite-stimulating effect of wine. The current line-up of suspects includes leptin suppression and alterations in a variety of other neurotransmitters and hormones, including neuropeptide Y, serotonin, gamma aminobutyric acid, and glucagon-like peptide-1.[9]

## Alcohol Ignorance = Pig-Ignorance

Returning to the Standard Life study, 44 per cent of women and a third of men insisted they were very conscious about the number of calories they consume through food. Ten per cent of women reported that they counted the calories on a daily basis with their food intake; more than 60 per cent of women and three quarters of men reported that they either don't know how many calories are in alcohol or weren't sure.

What is interesting and relevant is the pattern of drinking which produces all the excess calories. In the study, alcohol calorie consumption was at its highest between Friday and Sunday with 68 per cent of women and 64 per cent of men upping their intake as soon as work was over on Friday. Both men and women drink 25 per cent more between Friday and Sunday than they do between Monday and Thursday each week. And the excess calories seem to come from binge-style drinking.[10]

Epidemiologists at the US Department of Health and Human Services have found an interesting link between binge-type drinking and body fat. Drinking pattern consists of two components: the amount consumed on drinking days (quantity) and how often drinking days occur (frequency). Previous studies generally examined drinking based only on average quantity consumed over time. However, this provides a limited description of alcohol consumption as it does not account for drinking patterns. For example, an average quantity of seven drinks per week could be arrived at by having one drink each day or seven drinks on a single day.

The epidemiologists believe that the BMI of people who drink alcohol may be related to how much *and* how often they drink. In an analysis of data collected from more than 37,000 people, researchers found that BMI was associated with the number of drinks people had on the days they drank: 'In our study, men and women who drank the smallest quantity of alcohol – one drink per drinking day – with the greatest frequency – three to seven days per week – had the lowest BMIs, while those who infrequently consumed the greatest quantity had the highest BMIs.' In trying to interpret these findings the lead researcher explained, 'Alcohol is a significant source of calories, and drinking may stimulate eating, particularly in social settings. However, calories in liquids may fail to trigger the physiologic mechanism that produces the feeling of fullness. It is possible that, in the long term, frequent drinkers may compensate for energy derived from alcohol by eating less, but even infrequent alcohol-related overeating could lead to weight gain over time.' The scientists finished with these words: '... drinking patterns were strongly associated with BMI, a finding of potential public health significance.'[11]

In addition to its appetite-stimulating effects, alcohol disrupts fat metabolism. A Danish study in 2004, published in the *American Journal of Clinical Nutrition*, observed that drinking alcohol suppresses the number of fat calories your body burns for energy – far more so than meals rich in protein, carbohydrate or fat.[12]

## Recipe for a Beer Belly

Researchers are beginning to uncover the mystery of the so-called beer belly by looking at insulin resistance, insulin secretion and abdominal obesity – major predictors of type 2 diabetes, which is becoming increasingly prevalent among British children, while in the US, childhood diabetes seems out of control and Japan too is suffering badly. This trend bodes very badly for the health of this generation and is strongly linked to excess body fat.[13]

In assessing whether alcohol was helpful in preventing diabetes, researchers in the Swedish Uppsala Longitudinal cohort study (2007) found that alcohol intake in older men did not improve insulin sensitivity. They also said there was a very 'robust' association between alcohol intake, waist circumference and waist-to-hip ratio. They pointed out that a high alcohol intake, especially spirits, was closely associated with abdominal body fat, not just overall body mass.[14]

Abdominal fat accumulation is not just a cosmetic problem, it can be a serious health risk. Abdominal fat, also known as 'android' or 'central' obesity, increases the risk for cardiovascular disease, high blood pressure, high blood lipids, glucose intolerance and elevated insulin levels.

Hormones may also be strongly involved because high alcohol intake has been shown to decrease blood testosterone in males, and also increase cortisol levels, which can lead to abdominal fat accumulation.

## Beer Bosoms

Moving north above the waistline lie so-called 'man-boobs' (or 'gynaecomastia', to give them their official title).

In 2010, the British Association of Aesthetic Plastic Surgeons (BAAPS) announced, 'Britons over the moob: male breast reduction nearly doubles in 2009.' The BAAPS went on to explain: 'The most impressive stats have been recorded specifically in male surgery ... with the number of gynaecomastia (or man-boob) ops alone having shot up by 80 per cent.'[15] And the Health Faculty of Leeds Metropolitan University issued a public statement, 'Moobs in the news', announcing that one of their researchers was doing a PhD entitled, 'Bruises heal but moobs last for ever – men's account of cosmetic surgery for gynaecomastia.' BAAPS member Dalia Nield reported that anything up to a third of the men seeking breast reductions are simply obese. Consultant plastic surgeon Rajiv Grover, Secretary of BAAPS, responsible for the UK national audit of male cosmetic breast reduction surgery, said: 'Quite a few cases are caused by obesity, and we often say to men to look at their lifestyles before thinking about the scalpel.'[16]

In advising modern men on preventing enlargement of their breasts, the University of Southern California University Hospital states that it is important to remove known risk factors, which include 'avoiding excess alcohol consumption'.[17]

## Teenage Binge

Returning to the issue of the pattern of drinking, let's look at how it may affect the accumulation of body fat in children and young people.

The Royal College of Psychiatrists reports that around a third of fifteen- to sixteen-year-olds in the UK binge drink three or more times a month – more than in most other European countries.[18] A University of Washington study has found that people who began binge drinking at age thirteen and continued throughout adolescence were nearly four times more likely to be overweight or obese and almost three and a half times more likely to have high blood pressure when they were twenty-four years old than those who never or rarely drank heavily during adolescence. The study looked at the young adult health consequences of adolescent binge drinking (consuming five or more drinks on a single occasion) between the ages of thirteen and eighteen. The lead author of the study said, 'Young adults' history of binge drinking during the teenage years, irrespective of current levels of binge drinking, appears to have serious effects on their health by age twenty-four.'[19]

So the clear message here is to delay the introduction of alcohol and reduce the likelihood, frequency and severity of young binge drinking. And while children may fall asleep at the mention of liver disease, the good news about the link between alcohol and body fat is that the short-term aesthetic effects of alcohol does interact with their narcissism to attract their interest. To the average girl in particular – fat matters.

## Reaches the Parts Other Drinks Cannot Reach

Moving on again, below the growing moobs and paunch this time, alcohol is having effects south of the border that don't exactly involve enlargement, but rather the opposite.

While most lads are quite happy to swagger while showing off the broken ankle acquired on the slopes when drunk, curiously few are willing to share a far more common problem: alcohol-induced brewer's droop, now rebranded as erectile dysfunction or ED. And although a night of brewer's droop is temporary and we can soon put it behind us, scientists are now discovering longer-term and permanent effects of binge drinking on the parts we'd rather not 'fess up to, such as irreversible damage to the nerves in the penis and/or hardening of the arteries that supply it and disruption in the levels of testosterone and oestrogen which underlie sexual drive and function. And if that's not enough to make young men stand to attention, perhaps the finding that heavy drinking over a period of time can irreversibly destroy testicular cells, leaving men with shrunken testicles, will reach the parts that other information has failed to reach.

In 2010, a study entitled, 'Alcohol intake and cigarette smoking: Impact of two major lifestyle factors on male fertility', found that alcohol abuse apparently 'targets sperm morphology and sperm production. Progressive deterioration in semen quality is related to increasing quantity of alcohol intake.'[20]

Testosterone production is disturbed in chronic alcoholics and is affected by a variety of mechanisms as the result of alcohol. A study by the Department of Psychiatry and Neurology at the University of Göttingen in Germany warns, 'Little is known about the reversibility of these changes upon abstinence.' And regarding what happens even after people stop their drinking, they note, 'persisting disturbances of the hypothalamic–pituitary–gonadal axis in male alcoholics upon cessation of drinking'.[21]

Putting the lack of testosterone to the test in a different culture, the Family Planning Association of Hong Kong conducted the Men's Health Survey in 2004, involving 1506 men aged twenty to seventy. Entitled, 'The effect of alcohol drinking on erectile dysfunction in Chinese men', the study found that compared with men who don't drink, alcohol drinkers who consumed three or more standard drinks a week were two and a quarter times more likely to report erectile dysfunction.[22]

The NHS uses unambiguous language in describing the role alcohol plays in ensuring an uneventful evening down below. They warn that men may suffer from temporary impotence after drinking, in addition to which long-term heavy drinkers might also suffer from:

- loss of libido and impotence
- shrinking of the testicles
- reduction in penis size
- reduced sperm production
- loss of pubic and body hair
- enlargement of the breasts (as a complication of cirrhosis).[23]

Excessive alcohol drinking over a long period of time not only damages the liver but also raises the levels of triglycerides in the blood. It also leads to high blood pressure, heart failure and an increased calorie intake causing the arteries and vessels in the penis to be clogged up by harmful cholesterol resulting in erection difficulty. A study of 1709 men on 'modifiable risk factors and erectile dysfunction' carried out by the New England Research Institutes also warns: 'Mid-life changes may be too late to reverse the effects of smoking, obesity and alcohol consumption on erectile dysfunction. Early adoption of healthy lifestyles may be the best approach to reducing the burden of erectile dysfunction.'[24]

However, sensible drinking or even abstinence do not confer complete protection from brewer's droop – assertiveness classes may be required too; the *American Journal of Epidemiology* reported that 'new cases of erectile dysfunction are much more likely to occur among men who exhibit a submissive personality'.[25]

## Skinful, Skin-deep

Teenagers may care less about their long-term health than they do about their short-term appearance. So you may want to mention to them a study on acne in which alcohol was implicated. In 2002, the *Archives of Dermatology* (not quite as exciting reading for teens as *Heat* magazine) reported on a study comparing two remote cultures on different sides of the world – 'Absence of acne in two non-westernised populations'. The explorer-cum-dermatologists studied the complexions of the Kitavan people living on the Trobriand Islands near Papua New Guinea and the Aché hunter-gatherers of Paraguay. They observed that 'the astonishing difference in acne incidence rates between non-westernised and fully modernised societies cannot be solely attributed to genetic differences among populations, but likely results from differing environmental factors'. They went on to identify these factors, highlighting that 'the intake of dairy products, alcohol, coffee and tea was close to nil'.[26]

Other scientists have pointed to possible mechanisms involved in alcohol's influence on acne. Dermatologists at King's College Hospital in London reported that alcohol has 'a profound influence on immune function' and brings about changes in the small blood vessels in the skin. We all recognise a drunk person who has a flushed face and bright red nose – this is due to alcohol causing a widening of blood vessels in the skin. The dermatologists go on to mention that 'post-adolescent acne' may also be a sign of alcohol misuse. They also mentioned that 'the association between alcohol and skin disease is under-reported'.[27]

## Muscle Man

A report published in 2010 by the Advisory Council on the Misuse of Drugs (ACMD) suggests that anabolic steroids are increasingly being used by teenagers and men in their early twenties as a way to build muscle quickly. Professor Les Iversen, Chairman of the Council, said: 'It is becoming a big phenomenon in the UK. There is no question that the number [using the drug] for sporting reasons is now a minority ... The real growth has come in young users who want to improve their body image.' The steroids, which can be injected or taken in pill form, mimic the effects of the natural male hormone testosterone. Side effects such as gynaecomastia (male breast enlargement) are caused by the excessive levels of testosterone which are converted into the female sex hormone oestrogen.[28] And while fat may be a feminine issue, muscle is most definitely a masculine one: the numbers of females using anabolic steroids are relatively small, and there is a clear preponderance of male users.

Yet with such a strong desire for brawn, coupled with the marketing of alcohol as the drink of tough sportsmen, few boys or their parents have been made aware of the real effects of alcohol on future muscle. In its statement, 'Alcohol eats away at muscle mass', the American Council on Exercise, flexes its own:

> If increasing muscle mass is one of your goals, then think twice before you go out for a night of heavy drinking. Consuming alcohol in large quantities has a direct effect on your metabolism, causing fat to be stored instead of being utilised as an energy source. Alcohol contains seven 'empty' calories per gram, meaning that these calories don't provide you with any of the essential nutrients you need to build that muscle mass.[29]

Add to this the American College of Sports Medicine (ACSM) Position Statement on the Use of Alcohol in Sports, which emphasises that drinking alcohol can be detrimental to athletic performance and recommends that alcohol be avoided for at least forty-eight hours prior to an event: '… alcohol appears to have little or no beneficial effect on the metabolic and physiological responses to exercise. Further, in those studies reporting significant effects, the change appears to be detrimental to performance.'[30]

I often have to explain to adolescents at the schools I visit that if alcohol really did make you a tougher guy, boxers and 'cage fighters' would be hitting the bottle before climbing into the ring/cage to hit one another, and that when I speak to fighters many make it clear that they don't drink at all, while those who do drink very, very little and not very often. This approach seems to arrest adolescents' attention far more than earnest discussions about drinking and driving or cardiovascular risk probability tables.

## Vintage Looks

Teenagers are hardly concerned with wrinkles – they're too busy trying to 'die before I get old'. But perhaps their parents may be more worried when they hear that researchers have found that drinking damages part of the cells that are linked to premature ageing and cancer.

It's been discovered that alcohol causes stress and inflammation to telomeres – the ends of DNA strands that stop them unravelling, much like the ends of shoelaces. As people age, telomere lengths shorten progressively, until eventually they are so damaged the cell dies, and alcohol apparently accelerates this process. Since telomere shortening is thought also to increase cancer risk, the researchers speculated that those with shorter telomeres due to heavy alcohol consumption would have an increased risk of cancer.

Andrea Baccarelli, the lead researcher at the University of

Milan in Italy, said: 'Heavy alcohol users tend to look haggard and it is commonly thought heavy drinking leads to premature ageing and earlier onset of diseases of ageing.' If you're literally dying to resemble Keith Richards – drink up, drink up.[31]

## Epigenetics

There is increasing interest in the way that alcohol causes 'epigenetic' changes in humans – that is influencing our DNA and changing the way in which our genes function – even before birth, and how this may change behaviour and health too. For example, one study in the *Journal of Neuroscience* stated that, 'Drinking alcohol causes widespread alterations in gene expression that can result in long-term physiological changes.'

A number of excellent studies have identified a variety of genes that are up- or downregulated by short- or long-term exposure to alcohol in experimental animals and humans. The researchers have actually identified one of the molecular mechanisms responsible, which they believe could play an important role in a variety of damaging physiological changes triggered by drinking alcohol.[32]

## *Vive la Différence?*

Consumed in moderation, alcohol is thought to help reduce the risk of heart disease. Indeed, alcohol consumption in conjunction with high intakes of fruit and vegetables may well explain the so-called 'French paradox'. The French diet is considered to be very high in fat, especially saturated fat, yet the death rate from coronary heart disease (CHD) seems lower than one would expect. Plausible reasons for this include the role of alcohol in increasing the 'good' HDL cholesterol, preventing blood plasma from becoming too thick and reducing inflammation.[33] Red wine, in particular, also contains flavonoids that act as antioxidants, thought to help reduce the build-up of atherosclerosis (when fat builds up on the inner walls of arteries). It is also claimed that red wine seems to help maintain the flexibility of the blood-vessel walls.

Writing in the *International Journal of Wine Research*, Dr Giuseppe Lippi and colleagues claim: 'The healthful and nutritive properties of wine have been acknowledged for thousands of years, but the observation that moderate consumption of red wine on a regular basis may be preventative against coronary disease is recent ... These effects are attributable to the synergic properties of several biochemical components of wine (alcohol, resveratrol, and especially polyphenolic compounds), particularly the red varieties.'[34]

Yet, despite the comforting headlines there is disagreement about the health benefits of moderate drinking. A research team in Sweden looked at hard economic facts as to whether low levels of alcohol actually make people healthier by keeping them out of hospital or off sick from work. And the news for believers in modest drinking is very disappointing: 'Low alcohol consumption carries a net cost for medical care ... Low alcohol consumption also causes more episodes in medical care. The only age group, for both genders, that show a protective effect of low alcohol consumption on medical care costs are eighty plus.'[35]

A decade ago, epidemiologists began to question the enthusiastically embraced assumption that red wine or alcohol in general extends your life. For example, Dr Ian Graham, a professor of epidemiology at Trinity College in Dublin, suggested that the lower rate of coronary deaths in France could be due to 'competing causes of death' – that is, that many more French men might die early from alcohol-related causes before they have the opportunity to die of heart disease.

It's rather a case of Give Heart Disease a Chance.

The world view that the French are able to control their drinking habits is untrue. Even in 2003, a seminal study published in *European Addiction Research* said that 'Alcohol is the drug that gives rise to the greatest cost in France' – more than all other drugs combined and tobacco too. One of the researchers – Pierre Kopp, professor of economics at the Sorbonne – commented: 'There is a collective misunderstanding of the dangers of alcohol in a country where a regular intake is claimed as a protection against heart problems ... Consumption is exceptionally high and the final bill is extremely heavy.'[36]

Getting back to alcohol and death, while the media focuses on alcohol preventing death from heart disease they – and we – would prefer not to think that death (and premature death) comes from many different quarters, heart disease being only one. So while it may be that some older men may gain a benefit of reduced cardiovascular risk from drinking a very small amount of wine, they could die prematurely from something else.

In a report for the European Commission, the Institute for Alcohol Studies pronounced as follows on the death prevention v. health-giving beliefs of modest drinking issue:

> There is a positive, largely linear relationship between alcohol consumption and risk of death in populations or groups with low coronary heart disease rates (which includes younger people everywhere) ... Thus, the overall risk of death is a balance between the harms that alcohol causes, which can be present at all ages, and the benefits from coronary heart disease, which is largely an illness in older age. This means that for women under the age of forty-five years and for men under the age of thirty-five years, *the level of alcohol consumption with the lowest risk to death is zero* [author's emphasis]. In very old age, it seems that the reduced risk for coronary heart disease is much less, and it is likely again that any level of alcohol consumption might increase the risk of death.

And the girls won't like to hear this: 'At any given level of alcohol consumption, women appear to be at increased risk from the chronic harms done by alcohol, with differing sizes of risk with different illnesses.'[37]

Most importantly, when headlines tell us that light drinkers live longer than non-drinkers, journalists don't ask the obvious question: could it be that there is something else about non-drinkers or their lifestyles, independent of their non-drinking, that may explain the real reason why they may not live as long as light drinkers. And the answer is 'yes' – non-drinkers frequently do not represent a normal healthy control group with which to compare light drinkers.

In the above report for the European Commission, the IAS identified 'unhealthier lifestyle factors in abstainers', such as being older and non-white,[38] being widowed or never married, having less education and income, lacking access to health care or preventive health services, having multiple health conditions (such as diabetes and hypertension), having lower levels of mental wellbeing, being more likely to require medical equipment, having worse general health and having a higher risk for cardiovascular disease.[39] And an Australian study found that non-drinkers had a range of characteristics known to be associated with anxiety, depression and other facets of ill health, such as low-status occupations, poor education, current financial hardship, poor social support and recent stressful life events, as well as increased risk of depression, all of which could explain an increased risk of heart disease compared with light drinkers.[40, 41]

Common sense should tell us that if it were really true that drinking alcohol was better for our health than not, governments and medical schools would be advising non-drinkers to take up modest drinking of red wine so as to improve their health and live longer. However, they are certainly not advising non-drinkers to 'make an appointment with your GP to discuss how you too can take up drinking'. In fact, in 2009, the President of the Royal College of Physicians told the

House of Commons in formal evidence that '... in population terms, there is a linear relationship between any consumption and harm'.[42] This should tell us all something.

We all want to hear and believe that it is actually better to eat chocolate or drink wine because we enjoy them – and yet some of us suspect that they're actually not benefitting our health. Instead of trying to justify our pleasures, possibly by misinterpreting the studies and believing the headlines, we'd be better off simply enjoying alcohol without holding it up as a health-food elixir. If we, as adults, consider alcohol for ourselves in an honest light, society's attitude towards young drinking will, in turn, change for the better.

## Seven

# Cheers! Let's Drink to Your Mental Health

Depression is not merely a feeling. It's now clear that it has a brain chemistry and hormone dimension to it: changes in our brain chemistry or hormones can affect our feelings, including those of happiness or depression. Since alcohol has a direct impact on the brain, and it also alters a person's biochemistry, researchers have begun to ask the obvious question: can binge drinking trigger depression?

The assumption that the relationship between depression and alcohol only runs in one direction – that is, that depression leads to drinking – is being turned on its head. New studies have concluded that heavy drinking can *cause* major clinical depression. It has also been found that both minor and major depression are primarily related not to how often we drink, but how much we drink per occasion, that is binge drinking.

## Half Full or Half Empty?

Although researchers have, in the past, identified a link between alcohol abuse or dependence and major depression, it hasn't been clear whether one disorder causes the other, or whether a common genetic or social/lifestyle factor increases the risk for both conditions.

A study in 2009 at the Christchurch School of Medicine and Health Sciences, New Zealand, included 1055 teenagers and young adults born in 1977 who were followed from birth (presumably a time when turning to the bottle meant something quite innocent involving a teat and calcium) and monitored to ensure they had no major physical or emotional problems. They were then assessed for alcohol abuse or dependence and depression at ages 17–18, 20–21 and 24–25. At 17–18, 19.4 and 18.2 per cent of them respectively met criteria for alcohol problems and major depression; at ages 20–21 the figures were 22.4 and 18.2 per cent; and at ages 24–25, 13.6 and 13.8 per cent.

The researchers found that at *all* ages, alcohol abuse or dependence was associated with a 90 per cent increased risk of major depression: 'The findings suggest that ... problems with alcohol led to increased risk of major depression, as opposed to a self-medication model in which major depression led to increased risk of alcohol abuse disorders.' Most extraordinary was that they pointed to a one-way route from drink to depression: 'a unidirectional association from alcohol abuse or dependence to major depression, but no reverse effect from major depression to alcohol abuse or dependence'.

According to the researchers, the underlying mechanisms that give rise to such an association are unclear; however, this link may arise from genetic processes in which the use of alcohol acts to 'trigger genetic markers' that increase the risk of major depression. In addition, alcohol's depressant characteristics may lead to periods of depression 'among those with alcohol abuse or dependence'. Alcohol abuse may also cause social, financial and legal problems that cause stress and increase the risk of depression, they said.[1]

In 2010, a further study, examining the link between alcohol and depression, published in *Social Psychiatry and Psychiatric Epidemiology,* also mentions 'direct causal pathways from substance use to depression ... in most cases, the most plausible explanation of causality is the one in which substance-use disorder increases the risk of internalising disorder [depression]'.[2]

Serotonin, which is manufactured in the brain, acts as a neurotransmitter – a type of chemical that helps relay signals from one area of the brain to another. There are many researchers who believe that an imbalance in serotonin levels may influence our mood in a way that leads to depression. Possible problems include: low brain-cell production of serotonin, a lack of brain-cell receptor sites able to receive the serotonin that is made, serotonin's inability to reach the brain-cell receptor sites or a shortage in tryptophan, the chemical from which serotonin is made. If any of these biochemical glitches occurs, researchers believe it can lead to depression.

And there is now strong evidence to suggest that binge drinking alcohol alters the function of the brain cells which use serotonin (5-HT3 receptor), possibly through actions on its receptor protein.[3] In a 2007 study, it was shown that repeated alcohol consumption in mice (5 per cent alcohol – the equivalent of mild lager – every third day for eighteen days for increasing time periods) deregulates serotonin function within the nucleus accumbens (a key structure in the brain implicated in the neurobiology of major depression) by reducing the concentration of serotonin outside of the cells.[4]

In 2001, researchers were picking up on binge drinking being associated with a raised risk of depression in women.[5] More recently, a study of 14,063 Canadians aged eighteen and over looked for two quite different types of depression: 'meeting criteria for a clinical diagnosis of major depression and recent depressed feelings' and how they related to different aspects of drinking, especially the amount and the pattern.

The good news here for parents is that depression is primarily related to drinking larger quantities per occasion, and is unrelated to drinking frequency: 'Depression is most strongly related to a pattern of binge drinking,' said the lead investigator. 'A pattern of frequent but low-quantity drinking is not associated with depression.'

Also, the overall relationship between depression and alcohol consumption was found to be stronger for women than for men, but only when depression is measured as meeting a clinical diagnosis of major depression. 'This study underscores the important fact that women and men differ in significant ways – both biologically and socially – that may impact how they drink and the predictors and consequences of their drinking behaviour.'

Those of you who feel encouraged to drink for your mental health may be disappointed to hear 'there was no evidence that light drinking was protective for major depression when compared with lifetime abstainers'.[6]

Of course, many of us will say that a drink actually makes us feel happy. But for some people, alcohol is, in the short term, 'biphasic' in its effects, initially producing a sense of euphoria which then turns to feelings of depression as the blood-alcohol levels fall. And alcohol in large amounts, whether taken to treat depression or not, produces a depressant effect on mood.

## The Road to Remorse

Generally, it seems that alcohol can work in one of two ways: you regularly drink too much (including 'binge drinking'), which makes you feel depressed, or you drink to relieve anxiety or depression. Either way, as we've seen, alcohol affects the chemistry of the brain, increasing the risk of depression.

Hangovers can create a cycle of waking up feeling ill, anxious, jittery and even guilty. Life becomes more depressing, and arguments with family or friends, trouble at work, memory and sexual problems may ensue.

The Royal College of Psychiatrists states that 'alcohol seems to have the same depressant effect in younger people as it does in adults'.[7] However, there is one big difference: as discussed previously (see p. 83), young brains are still developing until the age of twenty-five, and if alcohol is leading to depression during this key developmental period, it is far more likely that it will have profound long-term effects. Symptoms of depression in adolescence are a strong predictor of an episode of major depression in adulthood, even among adolescents without major depression.[8] And the earlier the age at which an episode of depression is experienced, the more likely it is to be long-lasting if untreated. Diagnosing and treating children and adolescents with depression is critical in preventing harm to their academic, social, emotional and behavioural functioning and in allowing them to live up to their full potential.

## Suicide and Self-harm

The Royal College of Psychiatrists asks the question: why is an alcohol problem together with depression a particular worry? And then answers it fairly succinctly: 'Regular heavy drinking makes you more likely to get depressed – and, indeed, to kill yourself.'[9]

Around a third of young suicide victims have drunk alcohol before their death, and increased drinking may have been to blame for rising rates of teenage male suicide.[10] And a much higher incidence of suicide in general, both completed and attempted, is associated with alcohol; the lack of self-control, compromised judgment and impulsivity that it generates can increase the chances of a person attempting suicide.

And some of the physical manifestations of depression are getting more sinister: youth agencies report anecdotal evidence that teenagers who once used a penknife on their wrist now cut or burn themselves secretly and badly. The number of children aged ten to eighteen years old admitted to hospital due to self-harm has risen by a third in five years, according to National Health Service figures,[11] while a report in 2010 by the Royal College of Psychiatrists emphasises that the rate of self-harm among young people in the UK is one of the highest in Europe. And, they believe, 'this greatly underestimates the problem since many people do not attend hospital'.[12]

Enter alcohol.

A study in 2007 by the NHS entitled, 'Harmful Drinking: Alcohol and Self-harm', found that problems relating to alcohol use are common among self-harm patients. Of the 3004 patients who ended up at emergency departments following an episode of self-harm, alcohol was cited as a contributory factor in 40 per cent of cases, while the victims themselves revealed an even stronger link: 'Sixty-two per cent of males and 50 per cent of females who attended an emergency department following self-harm reported consuming alcohol ... 27 per cent of men and 19 per cent of women cited alcohol as the reason for self-harming.'[13]

## Younger Depression and Drinking

The chief executive of the mental health charity Sane has said what many others are observing in many industrialised countries: 'We're getting clear evidence that the onset of depression is happening earlier and earlier ... In previous generations, people would be overwhelmed by depression in their twenties. Now the peak age for onset is thirteen to fifteen: the numbers of teenagers calling us for help suggest the rates of depression in the under-fourteens have doubled in the last four years, and in the fifteen to twenty-four age group it has increased by one third.'[14]

In line with this rise, the use of antidepressants for children is also on the increase. More than 113,000 prescriptions of antidepressants were issued to children under sixteen in 2007 alone, and nearly 108,000 to sixteen- to eighteen-year-olds.[15]

The road between alcohol and depression may run in either direction; in some cases, alcohol misuse can lead to depression, while in others, depression can lead to alcohol misuse. Unlike physical diseases, which can be observed and measured, depression – like most psychological illnesses – is complex and difficult to diagnose because the mind is abstract and invisible. Yet the media, the legal system and, of course, the pharmaceutical companies, thrive on neat categories and labels to help make sense of the murky world of feelings and the mind. And we too feel a need to understand intangible feelings, and are reassured if they can be identified, described and prescribed for.

To give you a general feel for the range of symptoms, here are some of the typical signs of clinical depression listed by the American Psychiatric Association, titled 'Symptoms of Major Depressive Disorder Common to Adults, Children, and Adolescents'. Five or more of these symptoms must persist for two or more weeks before a diagnosis of major depression is indicated:[16]

- Persistent sad or irritable mood
- Loss of interest in activities once enjoyed
- Significant change in appetite or body weight
- Difficulty sleeping or oversleeping
- Psychomotor agitation or retardation
- Loss of energy
- Feelings of worthlessness or inappropriate guilt
- Difficulty concentrating
- Recurrent thoughts of death or suicide.

But if depression in general is difficult to identify, in children and adolescents it is more difficult still. Although the 'diagnostic criteria' and key defining features of major depression in children and adolescents are similar to those for adults, recognition and diagnosis of the disorder is complicated by the fact that the way symptoms are expressed varies with the developmental stage of the child. In addition, children and young adolescents with depression may have difficulty in properly identifying and describing their internal emotional or mood states. For example, instead of telling us how bad they feel, they may act out and be irritable towards others, which may be interpreted simply as bad behaviour or disobedience. Many children also come to hold a paralysing belief that there is no label or solution for the painful feelings which are blighting their lives.

Signs that may be associated with depression in children and adolescents are:

- Frequent vague, non-specific physical complaints such as headaches, muscle aches, stomach aches or tiredness
- Frequent absences from school or poor performance
- Talk of or efforts to run away from home
- Outbursts of shouting, complaining, unexplained irritability, or crying
- Being bored
- Lack of interest in playing with friends
- Alcohol or substance abuse
- Social isolation, poor communication
- Fear of death
- Extreme sensitivity to rejection or failure
- Increased irritability, anger or hostility
- Reckless behaviour
- Difficulty with relationships.

Some research shows that parents are even less likely to identify major depression in their adolescents than are the adolescents themselves.[17] Furthermore, many adolescents are not always the most reliable or even willing reporters of their own symptoms.

While an adult is more likely ultimately to see a doctor, adolescents and certainly children almost never do until their symptoms are very strong. It therefore falls to adults to spot the symptoms of depression early on. However, few parents – understandably –

want to admit their child has a serious mental-health problem, not just because of what it might mean for their child, but also because of what they think it might imply about them as parents. They prefer instead to believe their child is going through 'normal' teenage angst.

Even if we do recognise that a child is depressed, treating the condition in children and adolescents remains a challenge. Few studies have established the safety and effectiveness of treatments for depression in the young.

## What Causes Depression?

Depression is usually caused by a mixture of things, rather than any one thing alone. Obviously, events or personal experiences can be a trigger, including family breakdown, the death or loss of someone you love, neglect, abuse, bullying and physical illness. Depression can also be set off by too many life changes happening too quickly. People are more at risk of becoming depressed if they are under a lot of stress, have no one to share their worries with and lack practical support. And, as with a number of psychological problems, there are biological factors too. Depression may run in families due to genetic factors. It is also more common in girls and women compared to boys.

As with our everyday feelings of low mood, there will sometimes be an obvious reason for becoming depressed, sometimes not. It can be a disappointment, a frustration, or the loss of something – or someone – important to you. There is often more than one reason, and these will be different for different people; according to the Royal College of Psychiatrists, these are some of them:

- Events
- Circumstances
- Physical illness
- Alcohol
- Gender
- Genes
- Society.

## Events

It is normal to feel depressed after a distressing event, such as a bereavement or divorce. A person may well spend a lot of time over the weeks or months that follow, thinking and talking about the event in question. After a while, they come to terms with what's happened. But they may get stuck in a depressed mood, which doesn't seem to lift.

## Circumstances

If someone is alone, has no friends around, is stressed, has other worries or is physically run-down, they are more likely to become depressed. Consider if your child is being bullied at school? Does she have a learning disability, which means that she is underperforming, leading to decreased self-esteem and depression? Have there been any changes at home, such as a death in the family, recent move, divorce, etc.? How come she doesn't have any friends?

## Physical illness

Physical illnesses can affect the way the brain works and so cause depression. They include:

- Life-threatening illnesses like cancer and heart disease
- Long and/or painful illnesses, like arthritis
- Viral infections like flu or glandular fever – particularly in younger people
- Hormonal problems, like an underactive thyroid.

## Alcohol

Regular heavy drinking is more likely to lead to depression – and, indeed, even suicide.

## Gender

Women seem to get depressed more often than men. It may be that men are less likely to talk about their feelings and more likely to deal with them by drinking heavily or becoming aggressive. Women are more likely to have the double stress of having to work and look after children.

### Genes

Depression can run in families. Where one parent has become severely depressed, their child is about eight times more likely to become depressed as well.[18]

### Society

The tide of depression today is seen as the result of social changes, particularly in 'developing countries' – the increasing exposure to global media which, in turn, has been found to separate people within communities and erode moral certainties. Looking more carefully at some of these big social changes is illuminating.

## Divorce/Separation

Across time and across cultures, family disruption has been regarded as something that threatens a child's wellbeing and even survival. But unlike bereavement or other stressful events, it is almost unique to divorcing or separating families that as children experience this significant change, the usual and customary source of support – parents/family – tends to dissolve.[19]

While we may look for signs of unhappiness in our children during or shortly after a divorce, there is often a 'sleeper effect' whereby the child exhibits effects years, even decades later. For example, a Canadian study currently in press has found that children whose parents divorced grow up to become adults with twice the risk of having a stroke. The link cannot be explained by other health factors and practices linked to stroke risk.[20] Furthermore, children of broken families are twice as likely to become depressed as adults. One study, 'The Long Reach of Divorce', shows how the effects are passed on to grandchildren – even those who haven't yet been born.[21]

Most of us accept that in adults divorce may lead to heavier drinking, and so it's not inconceivable that divorce could lead children to drink more. A survey commissioned by a UK legal firm, of 2000 individuals who had experienced their parents' divorce as a child in the past twenty years, was reported to have found that a quarter of children whose parents divorce before they reach eighteen 'turn to alcohol'.[22]

As a society with one of the highest divorce rates in Europe, we in Britain may be forgiven for preferring to believe that kids 'bounce back' after a divorce/separation. We prefer research which reinforces, or at least does not contradict, our lifestyle. So rather than face up to the fact that divorce may come with a sense of failure that we must carry, we prefer to watch programmes or acknowledge studies that show that divorce doesn't affect children as much as we thought. However, like a developing flower, children need the right type and amount of nurturing, especially during and in the wake of their parents' divorce, or they will wilt. This is much more important than our guilt.

## Preventing Depression

Children need our time and availability to listen as an outlet enabling them to process and integrate their emotions. They also need their own time to turn inward, as opposed to the unlimited outward distractions on offer, to enable some emotional housekeeping and general psychological maintenance to take place. As our society becomes more time-poor and technology-rich this isn't happening enough. With teenagers it's a case of respecting their need for more privacy while reducing their isolation.

To put it plainly, face-to-face time between family members produces biological changes that translate into better relationships and happier children. We must try to communicate regularly with our children. Also we should automatically be enabling an ongoing form of soft monitoring to see if a child is lethargic or antisocial and encourage them to keep communication open.

A study at Columbia University reports that having at least one parent eat dinner with their child regularly was found to prevent depression, anxiety and substance abuse in children, who also achieve higher grades in school, compared to those children who dine on their own.[23] Eating together has also been cited as strengthening marital relationships and children's 'self-esteem'.[24] Children in families who eat together fewer than three times per week reported higher levels of family tension, less conversation and lower self-esteem than those who eat together more often, without television.[25]

Research at the National Center on Addiction and Substance

Abuse at Columbia University 'consistently finds that the more often children eat dinner with their families, the less likely they are to smoke, drink or use drugs'. The National Center on Addiction and Substance Abuse has established 'Family Day – a Day to Eat Dinner with Your Children; eating dinner frequently with your children and teens reduces their risk of substance abuse'. The chairman has even gone so far as to state: 'One of the simplest and most effective ways for parents to be engaged in their teens' lives is by having frequent family dinners ... one factor that does more to reduce teens' substance abuse risk than almost any other is parental engagement.'[26] And the US Congress now recognises National Family Month (mid-May to mid-June), strongly promoting 'frequent family meals' as the way 'to help America's families reconnect'.[27]

Finding creative expressive outlets for our children to engage in is another powerful way of coping with and processing emotions and will be invaluable to a child who might be at risk of developing depression.

## Sleep, Depression and Drinking

Sleep deprivation in children and young people (not to mention their parents) is a growing problem with a wide range of effects, including moodiness and depression.

A study published in *Current Biology* in 2007 concluded that sleep is essential for emotional regulation and processing and that 'even healthy people's brains mimic certain pathological psychiatric patterns when deprived of sleep'. In healthy, young people deprived of sleep the researchers found that 'the emotional centres of the brain were over 60 per cent more reactive under conditions of sleep deprivation than in subjects who had obtained a normal night of sleep'.[28]

Aside from the link between sleep deprivation, depression and drinking, there is a more direct link between sleep deprivation and heavier drinking in young people. Some researchers believe that variations in the duration of night-time sleep and the level of daytime sleepiness may play an important role in influencing alcohol consumption. A British survey found shorter periods of sleep were associated with heavier drinking. Similarly, an American study of young adults found that those needing only six hours of sleep or less had an earlier age of drinking onset and drank more per month than did those who needed more sleep, leading the investigators to hypothesise that short sleep is associated with heavier drinking. 'Sleep quality and daytime sleepiness may also relate to rates of alcohol drinking and become a gateway to excessive alcohol use.'[29]

## Physical Activity and Depression

In 2008, one hundred specialists in exercise and neurobiology attended a two-day conference to explore the potential for physical activity to prevent substance abuse. The conference announced $4 million in new research grants to help. There are some tantalising clues that physical activity might spur changes in the brain to prevent depression, and now the US government is beginning a push for hard research to prove it. This is not about the so-called runner's high, which is experienced after fairly intensive exercise.

Instead, the question is just how regular physical activity of varying intensity – dancing, cycling, swimming, football, skateboarding – might affect mood and even the very reward systems in the brain that can be hijacked by alcohol abuse.[30]

A review of 106 studies on fitness and health in children found more than the mere physical benefits doctors were looking for, concluding: 'Improvements in cardiorespiratory fitness have short-term and long-term positive effects on depression, anxiety, mood status and self-esteem in young people, being also associated with a higher academic performance.'[31]

At exactly the same time, two British studies were revealing a gigantic missed opportunity staring us in the face. Monitoring *actual* levels of physical activity by using portable recording devices, both studies found that only 2.5 per cent of our school-age children achieve the current (very modest) recommended daily levels of physical activity. Worse yet, these studies expose the fact that children's physical activity levels have been drastically overestimated, with true levels likely to be around six times lower than official government data which relies on information supplied by parents.[32]

As countries become more technological and Westernised, children's cardiorespiratory fitness (how efficiently their heart, lungs and other parts of the body deliver oxygen to muscle tissue) is declining by 4.3 per cent per decade globally. And new research has found that children's fitness levels in England are falling faster than anywhere else in the world. In a running test, the average ten-year-old from 1998 would beat 95 per cent of their peers today. The scientists describe this fall in fitness as 'large and worrying' and have no hesitation in blaming modern life: 'Children are not doing as much physical activities as before. Britain is highly up to date with technology, with more computers than anywhere else in Europe ... they are using their spare time to play more computer games, more time watching TV or more time online. They don't climb trees any more, they don't use their bikes any more.'[33] And while obesity and fitness are linked, in a lot of cases fitness levels in children of a healthy weight are poor too.

The sedentary child or teen turns into the sedentary adult. It is extremely important to establish good fitness levels and habits in children as early as possible because this influences their long-term fitness and health. Fitness doesn't necessarily mean formal classes though. As children naturally want to move around, it's rather a case of letting them do more of what comes naturally. It is the ambient background activity, such as spontaneous running, walking, lifting, playground fun and games, that really adds up to optimal physical fitness. And remember: vigorous is much better than moderate. Walking to school or accompanying a parent when they walk the dog all adds up.

Schools, in conjunction with parents, can easily devise ways of increasing physical activity, formally and informally, in and out of school time. We should never worry that physical activity will displace time that could be spent learning. The great Roman philosopher Seneca realised this two thousand years ago when he wrote, '*Roga bonam mentem, bonam valitudinem animi, deinde tunc corporis*,' meaning, 'Exercise promotes a good mind, good spiritual health, finally, health of body.'

The signs of the preventative approach are good. A twelve-year study of 14,000 people without any mental illness, published in the *Journal of Psychiatric Research*, found a striking dose–response relationship between level of fitness and the prevention of depression.[34]

In addition, beyond all of this, there is the issue of children not getting enough daylight outdoors. Sunlight will keep sleeping patterns stable and is absolutely vital to a child's development and health.

Given that there are more children and young people suffering from depression than ever before, ensuring that alcohol neither exacerbates the existing high levels of depression or causes more cases is of paramount importance.

Emphasising to our children that binge drinking may cause brain changes that could make them depressed is a good start. And helping them to see the glass as half full as opposed to half empty is more likely to keep them away from the bottle.

## Eight

# Accidental Damage

Both the stifling culture of Health and Safety itself and the backlash against it have overlooked a common thread in harm through accidents and victimhood – namely alcohol, the tonic of misadventure.

One of the most successful health-and-safety campaigns ever – even among young people – has been the drink-and-drive one. And yet, it's as if society only sees alcohol-related mishaps as something that happens behind the wheel of a car. If we accept that alcohol can impair reaction time, co-ordination, judgment and impulse control when driving, surely it should be obvious that these effects are equally applicable to every other area of our lives. Whether we're talking about car crashes, gap-year deaths or being a victim of crime (or even an unlikely perpetrator) – we need to take a step back and look at alcohol as a key player in many of the accidents involving teenagers and young people, as it meets with a brain that is desperately trying to reconcile urges and impulses with self-control.

It's worth surveying the range of mishaps that can result from a journey of misfortune guided by a beer.

## Pedestrian Matters

The necessary and beneficial focus of attention on drivers who drink and kill has diverted attention away from walkers who drink and die. Society is interested in breathalysing or, if they are already dead, testing the blood-alcohol levels (BACs) of drivers who are involved in accidents. However, by testing those on the other side of the windscreen a whole new picture emerges, since a huge proportion of pedestrians killed have also been drinking. According to Coroners and Procurators Fiscal in Britain, '74 per cent of pedestrians killed between 10 p.m. and 4 a.m. were over the legal limit for drivers'.[1]

Most New Year partygoers know better than to drive while they're drunk. What they probably don't know, however, according to the body counters (actuaries and epidemiologists), is that *walking* – even when BACs are at the legal limit – may be even more dangerous.

In the United States, 1 January is a deadly day for pedestrians, consistently averaging the highest number of walkers killed in motor-vehicle crashes and a greater proportion of on-foot victims who are 'legally drunk'. Alcohol impairment is being cited as the cause, with more pedestrian crash deaths than any other day of the year. Halloween, another drink-related holiday, at least for adults, comes a close second.[2]

But A&E staff say that the risks for drunk pedestrians are not only concerned with the likelihood of being run over by a car. One trauma surgeon has pointed out that while most drunk walkers are trying to do the right thing by parking their cars, they forget just how strongly alcohol can still affect them. 'The drunk walkers turn everyday objects into deadly weapons; they trip over a crack in the sidewalk, they hit their heads on curbs and light poles.'

Alcohol not only impairs co-ordination and the ability to walk, it also impairs judgment. And that lack of judgment may be why pedestrians are more often at fault when they're involved in fatal crashes. According to a 2002 study 'Pedestrian Crashes in Washington, DC and Baltimore', comparing pedestrian versus driver culpability, it was the *pedestrians* who were found to be far more likely to be responsible.[3] And few spare a thought for the *sober* driver who hits and kills a drunk pedestrian. There is no 'compensation' for them, no easy way to wipe the post-traumatic stress visions from their mind at night.

Rogue economists who have done the maths now claim that drunk walking may actually be deadlier than drunk driving: on a per-mile basis, a drunk walker is eight times more likely to get killed than a drunk driver, and walking drunk leads to five times as many deaths per mile as driving drunk.[4]

But good news has come from the *Journal of the International Society for Child and Adolescent Injury Prevention*:

pedestrians who drink less, die less. A nine-year study of 'changes in traffic crash mortality rates attributed to alcohol use by drivers and pedestrians' reported that decreased alcohol use among pedestrians was associated with substantial reductions in crash mortality.[5]

## Brewer's Bicycle

There isn't much data from Britain regarding BACs of dead cyclists. The Department for Transport explains: 'For many drivers or riders killed in road accidents, a post-mortem blood-alcohol level is not available [We apologise for any inconvenience caused by the lack of data.]'[6]

But the data available elsewhere is equally applicable here, if not more so. A study published in the *Journal of the American Medical Association* states that 'elevated blood-alcohol concentrations (BACs) are found in about one-third of fatally injured bicyclists aged fifteen years or older'. In particular, cyclists whose BACs were at the legal driving limit or over (80mg per 100ml) had a 'twenty-fold heightened risk of fatal or serious injury'. But even those whose level was low – only a quarter of the legal limit (so 20mg per 100ml) and over – were 5.6 times more likely to be killed or seriously injured on their bike. It's important to realise that these statistics excluded teenagers under fifteen years old along with any victim who survived for six hours or more after the accident before dying (because BACs would have dropped). Furthermore, the statistics came from daytime cycling until

9 p.m. only and the researchers emphasise therefore that 'it is conceivable that the risk of bicycling injury attributable to alcohol use is actually greater than reported in this study'.[7]

Another study including people of *all* ages found that 24 per cent of cyclists killed in 2006 had BACs at or above the legal driving limit.[8] And 21 per cent of autopsies for New York City cyclists who died within three hours of their accidents detected alcohol in their bodies. 'It's something we have to call attention to,' said the Director of the Injury Epidemiology Unit at the New York City Health Department. 'To learn this is new for us. We want to get that information out there.'[9]

In Hungary, a similar picture emerges. In a study published in 1999, in the upbeat journal *Forensic Science International*, Hungarian traumatologists reported that at the time of their hospitalisation 37.9 per cent of injured cyclists and pedestrians in traffic accidents were under 'moderate influence' of alcohol.[10]

## Victimhood

Ben Kinsella was stabbed to death in London in 2008, causing an outcry over 'knife crime'. He was murdered in a bar by three attackers with whom he'd had no dealings or interaction. He was utterly blameless. In fact, the judges went out of their way to emphasise that 'he bore no responsibility whatsoever', describing an 'arrogant and unfeeling attack on someone who had done nothing', and condemning the killers for picking on 'an obviously younger and smaller lone victim'.[11]

The murder led to a series of anti-knife crime demonstrations and a review and change of UK knife-crime sentencing laws. Yet in all the anger and debate that surrounded the case, mention of alcohol was conspicuous by its absence. Society did not ask why a sixteen-year-old boy (and his sixteen-year-old friends) was in a 'brasserie bar' at 2 a.m. celebrating the end of his GCSE examinations. It seemed to be a cultural norm.

In the United States, the proportion of corpses testing positive for alcohol is between one third and two thirds. For example, in a study of 5000 homicides in Los Angeles, toxicology reports noted alcohol in half of the corpses, while in 30 per cent of them the blood-alcohol level was high enough to classify the victim as legally intoxicated at the time of their demise.

Forensic scientists describe a typical alcohol-related murder as a young male stabbed to death in a bar on a Friday or Saturday night, following an argument with an acquaintance or friend. They even have theories of how the death comes about. For example, the 'selective-disinhibition view', whereby if one person is drunk – or worse, both people – their judgment will be impaired, and each is more likely to misinterpret the other's intentions, cues and actions, reacting in a less restrained way as a result. Or the 'outlet-attractor view', whereby at certain places (pubs, off-licences, clubs) young people gather expecting to let go of the normal constraints of daily life and act differently in that 'anything goes' environment.[12]

The Alcohol Research Consortium in the UK has an interesting way of measuring alcohol's role in violence among young people: 'Our ultimate aim is to try and reduce the incidence of facial trauma sustained as a result of alcohol and violence.' Just looking at the gruesome gaping and bleeding faces published in their papers and in medical journals, such as the *British Journal of Oral Maxillofacial Surgery*, gives one an immediate and graphic sense of what the headlines and statistics have been saying for a long time. The job of these surgeons is to put torn faces back together again when possible. But their cutting-edge industry has to contend with 'the complex link between alcohol and violence'.[13]

In order to protect them from harm sixteen- and seventeen-year-olds should *not* be in pubs, bars, clubs or 'brasserie bars', celebrating anything.

## So Here's to You, Mrs Robinson

As discussed in Chapter 5, the date-rape drug found in 'spiked' drinks has finally been identified and is now being referred to as alcohol. But rape and sexual assaults do occur in other contexts. Or, as a Government Equalities Office and Home Office independent review in 2010 put it, 'Rape is often part of something else.' And that 'something else' is all too often alcohol. The report by a woman, Baroness Stern, entitled 'The Stern Report', examined the treatment of rape victims by public authorities. And there were stern words indeed for alcohol, which featured prominently (I've put some key points in italics):

> Alcohol was discussed in almost every meeting we held. Senior staff at a Sexual Assault Referral Centre in the north of

England told us that *three quarters of adults who come to them have been drinking, and staff must wait for them to sober up* before seeking informed consent to the forensic examination. The police also told us of the impact of excessive alcohol consumption on those who come to them to report rape. We heard about the vulnerability brought about by excessive drinking, the difficulties of investigating cases when memories were clouded by drunkenness and the problem of taking such cases through the courts. We heard from many who spoke to us that the 'night-time economy' was a 'harm-producing arena' and 'very problematic', both in terms of general safety and in relation to the prevention of sexual assaults.

The report also noted that the Royal College of Physicians told the House of Commons Health Committee:

The 'passive effects' of alcohol misuse are catastrophic – rape, sexual assault, domestic and other violence, drunk driving and street disorder – alcohol affects thousands more innocent victims than passive smoking ... As we were told, 'rape is often part of something else'. Prescribing solutions is beyond the scope of our report, but if we did not acknowledge these concerns we would not be doing justice to many of those who spoke and wrote to us both about *the links between alcohol and rape* and about our *failure to protect some very vulnerable people from harm.*

Stern emphasised that many of the people they spoke to stressed the vital importance of teaching children and young adults about rape, sexual assault, consent and the role of alcohol. And, for boys, there are additional considerations: judges in England and Wales were told in late 2007 that a woman may not be capable of giving consent to sex if she is drunk, even if she is conscious and seemingly 'gagging for it'.

The rape trial in 2009 of chef and sociology student Peter Bacon, aged twenty-five, from Canterbury, Kent, and his accuser, a forty-four-year-old woman, was highly instructive. Bacon had sex at the woman's house after a night of heavy drinking. However, Winchester Crown Court was told that when the woman woke up the next morning, she started shouting and told Mr Bacon to leave. He told the court: 'She asked, "Did we have sex?" I said, "Yes," and she started shouting, "Rape!"'

Mr Bacon went on to say that the woman claimed she had been too drunk to consent and told him to get out.[14] He then went straight to the local police station and told them what the lawyer had said when he woke up in her bed. He explained that he wanted to find out what his legal position was. And the police were very obliging – he was arrested and later charged. The woman was still twice the drink–drive limit later that day when she had a medical examination.

In police interviews, Mr Bacon admitted that he knew the woman was drunk, but that 'she was still able to hold a conversation with me'. He said that he thought the woman had enjoyed the sex: 'She groaned. She gave the impression that she was enjoying it.'

The woman, who described herself as a 'recreational binge drinker', said she had no memory of having had sex and claimed she had been incapable of giving consent. She held that because she couldn't remember what happened, the sexual intercourse the pair had must have been non-consensual. In fact the Court of Appeal had come to a similar view in 2007. According to them, a woman who is drunk may well be unable to give her consent, but it is left ultimately to the jury to decide whether the man had a 'reasonable belief' that consent had been given.

Describing the legal issue of drunken consent as a 'vexed question', the judge said: 'If the complainant thought that the change in the law meant that the law no longer recognises that a drunken woman can give consent, she was completely wrong.' Sober, drunk or very drunk, he said, 'If a choice is exercised freely, then it is consent.' He continued: 'The fact that she doesn't remember what happened, doesn't prove that she didn't give consent, however bitterly she may have regretted it the next day.'

The jury of seven women and four men unanimously acquitted Bacon in only forty-five minutes. But seven months after he walked free, Bacon announced that he had been forced to change his identity by deed poll in order to start a new life abroad. Despite his acquittal, Bacon stated that he still felt that he was being punished. The fact that he was not given anonymity by the courts meant his name continued to be linked to a crime that never happened. His accuser, however, was automatically granted lifelong anonymity as an alleged rape victim.

The Stern Report tried to get to grips with the effect of alcohol on the ability to give consent:

> How far there can be consent to sex when one or both the people involved are very drunk has been a controversial matter. When the person complaining of rape is unconscious as a result of her voluntary consumption of alcohol, the starting point is to presume that she is not consenting to intercourse ... This is viewed by the Court of Appeal, as 'plain good sense'. The Court also said that if, through alcohol (or any other reason), the person has temporarily lost her capacity to choose whether to have intercourse, she is not consenting. If, on the other hand, despite having drunk a lot of alcohol, the person is still capable of saying yes or no and agrees to sex, then it is not rape. Whether or not the victim has voluntarily drunk too much does not lead to the conclusion that he or she voluntarily agreed to have sex.[15]

So in practice, what this means is that if your son and Miss X (or Mrs Robinson) are both drunk and she asks him to come to bed and have sex, she can, under current law, justifiably claim that she was not able to give proper consent and that it was rape. Before leaving office in 2010, the Labour government made it clear again that they want the conviction rate for rape to increase considerably.

## School Days

In June 2009, the Legal Alliance, an organisation of British legal firms, reported: 'A boarding school is being sued by a former pupil who was left disabled after falling from a window while drunk.'

Amy St Johnston was sixteen years old when, as a student at Oundle School in Northants, she fell 46 metres from her first-floor room during a Valentine's Ball in February 2005. Ms St Johnston, now aged twenty and confined to a wheelchair, has since left Oundle to study at Selwyn College, Cambridge, and is suing her former school for £300,000 in damages, claiming a breach of a duty of care for encouraging a 'drinking culture' among senior pupils.

A spokesman for the school, which allows sixth-formers to drink with 'a substantial meal', said legal advisers are handling the issue.[16]

Boarding schools offering pupils a Saturday night tipple, but also day schools holding school balls, have now to consider taking a position on offering their pupils alcohol or knowingly allowing pupils to drink. But this, I've discovered, is a patchy affair. When I visit British secondary schools, I note that some allow their pupils a couple of glasses of wine at age sixteen every Saturday night for fear that if they don't, the teenagers will get drunk elsewhere in town. Some schools breathalyse their pupils before they can have their weekly dose. Other schools don't allow this. Many top private schools find pupils unconscious, sometimes in the dark, in the snow. Parents and schools need to take some position on pupils and alcohol, but at the moment it's a somewhat vague position, or the parents take one, while the school doesn't or vice versa.

On the other side of the Atlantic, the *New York Daily News* reported in 2010 on a 'drunk and belligerent' seventeen-year-old Notre Dame football recruit – Matt James – who was killed in a fall from a fifth-floor hotel balcony during his senior-year spring break in Panama City Beach, Florida. He was dead when police arrived.

Local police reported, 'Witnesses and friends indicate he had become drunk and belligerent. He had leaned over the balcony rail, was shaking his finger at the people in the next room over. He fell over.' They said the railing at the hotel met the standards for proper height. The police announced they would be interested in pursuing charges if it was learned who provided the underage [under twenty-one] teen with alcohol. 'It apparently magically appeared,' they said. 'Nobody wants to tell us where it came from.'

Only a few days earlier, a nineteen-year-old died in similar circumstances. Police think alcohol was involved in that fall too. Panama City Beach Mayor Gayle Oberst said: 'It happens all over the country, and it's really a sad thing when it does happen ... Balcony falls are associated with spring break, I think, but they happen every year, all year, everywhere.' Oberst also talked about underage drinking and vacationers: 'Unfortunately, there's a lot of peer pressure, and then you get them away from home and they tend to drink too much and it almost always results, with a handful of people, in a tragedy for them,' she said.

Panama City Beach Police Major David Humphreys said he hoped other spring-breakers would take a lesson from James's death, but the scene at the beach the next morning appeared as if nothing at all had happened. Spring-breakers lay on the beach just a few metres from the patio where James had fallen and carted eighteen-packs of beer between their cars and their rooms. But, said Humphreys, 'We realise you're here to have a good time ... We ask that you don't do anything that you wouldn't do at home. And for those twenty-one or over, "Use alcohol in moderation."'[17]

In England, sixteen-year-old Paddy Higgins, of Winnersh, Berkshire, died on 6 July 2009 after falling down cliffs above Tolcarne Beach. He had been with friends and was celebrating the end of his GCSEs. Tests showed he was three times over the legal drink–drive limit. Higgins's death came just eight days after Andrew Curwell, eighteen, from Saddleworth, Lancashire, was found at the foot of cliffs at nearby Great Western Beach. The Leeds Rhinos' rugby academy player had been on holiday with friends to celebrate the end of his A-level exams when he fell from the cliffs.

A third teenager was found unconscious near the bottom of cliffs in the town in an unrelated incident. The sixteen-year-old had plummeted over 200 metres down a 300-metre drop in Newquay and it was thought he may have been there all night. Inspector Dave Meredith, of Newquay police, said: 'The police are appealing to all people visiting Newquay this summer to be aware of the dangers of going near cliff areas after consuming alcohol.' But even away from the cliffs another headline from Cornwall reported: 'Drunk teenager died falling from hotel balcony at works party, hears inquest.'[18]

So, yet again, it's not a case of drinking and driving – rather one of drinking and falling.

The Drinkaware Trust, the independent charitable trust aimed at promoting sensible drinking, funded by the alcohol industry, has shown increasing concern over this aspect of alcohol and the young, actually issuing a press release in 2010: 'Drinkaware warns parents to avoid giving their children alcohol over the summer.' It went on to warn that parents who provide sixteen- and seventeen-year-olds with alcohol to take on holiday may be inadvertently putting them at risk.

Drinkaware's survey found the following:

- Almost two fifths (39 per cent) of parents would give alcohol to adolescents going on a week-long holiday, if requested; of those, more than half would give five or more bottles of spirits or wine.
- More than a third (36 per cent) of parents would prefer their child to get alcohol from them, rather than from an unknown source.
- More than one in five said they buy their teenager alcohol in order to keep track of the amount they will drink.
- Parents are happy to give alcohol to their children, despite two-fifths knowing that their child has had a bad drinking experience.
- One in five parents is aware that their child has been involved in an accident or had unprotected sex when drinking, and 79 per cent know their child has been sick.

The chief executive of Drinkaware said, 'Parents might think they're doing the right thing by ensuring alcohol comes from them instead of somewhere else, but when young people drink to excess it can compromise their personal safety and increase the chances of them having unprotected sex or being involved in an accident.' He added: 'To help their children stay safe, parents should avoid giving them alcohol for unsupervised holidays.' And at the bottom of the press release they even mention: 'Alongside the Chief Medical Officer's guidance, Drinkaware advises that ideally all under-eighteens should enjoy an alcohol-free childhood.'[19]

## The New First Aid

As I write, the British Red Cross is reporting that one in seven eleven- to sixteen-year-olds has been in an emergency situation as a result of a friend drinking too much alcohol, according to a new study 'showing the vulnerability of young teens'. Of these, 532,128 children 'have been left to cope with a drunken friend who was sick, injured or unconscious. Half of these had to deal with someone who had passed out and a quarter had to deal with an injured friend who had been drunk and in a fight.'

Only one in ten of the young people rang 999 and less than half contacted their parents when faced with the responsibility of taking care of their friend. A large number of adolescents who stepped in to help were left distressed, unsure as to whether they had done the correct thing. Close to half were worried their friend would choke on their own vomit or wouldn't wake up.[20]

While we normally think of the Red Cross as helping out in foreign flood disasters, charity now begins at home. They announced, 'Teens feel ill-equipped to handle consequences of underage drinking', marking the launch of 'Life. Live it' – a British Red Cross campaign aimed at eleven- to sixteen-year-olds, to help them learn life-saving skills so that they are better able to cope in an emergency. In a comment on our times they said, 'Our aim is to make first aid accessible to young people and their everyday lives.' And drunkenness is, now, part of young teenagers' lives.

## Gap-Year Giddiness

There are no official statistics showing how many gap-year travellers are injured, attacked or killed while abroad. However, one in three gap-year adventures is cut short by an accident or crime, according to the organisers of a conference on the safety of young travellers held in London in 2009. Figures indicate a 10 per cent rise in numbers of people taking a year off to travel, but growing numbers of parents are contacting gap-year travel companies, concerned about their child's safety. The Gap Year Safety Conference had, in part, been called as a response to heightened parental concern and the number of gap-year tragedies.

The founder of GapAid, a charity established to promote safer travel for young people, said: 'We are not suggesting that you can avoid all risk, but there are so many issues which we don't think people are properly prepared for because there is no national curriculum for this.'[21]

The Foreign and Commonwealth Office Consular Directorate now offers travel advice at GoGapYear.com, including a section entitled, 'Be careful with alcohol'. It explains: 'When it comes to alcohol, make sure you know your limit. You're more likely to have an accident if you're drunk and probably won't be covered by your insurance.'

Again, as I continue to write, I'm interrupted by a report from Alcohol Concern, claiming that 'in the UK we have the highest rates in Europe of teenage alcohol-related injuries'.[22] And the *Nursing Times* and the *Health Service Journal* are both reporting: 'Alcohol-related hospital admissions soar.'[23]

Official statistics put the percentage of alcohol-related casualty admissions on Friday and Saturday evenings at around 20 per cent, a Commons health select committee was told. But in 2009, Robin Touquet, professor of emergency medicine at St Mary's Hospital in London, told MPs that a study had estimated the true figure to be around 70 per cent. Brian Hayes, from the London Ambulance Service, said paramedics have to deal with 'horrific injuries':

> We're talking about people that because of alcohol have jumped on a wall because of bravado in front of their mates, not realising the other side is a sixty-foot drop – and they've gone down it ... Their big massive night out has ended with a family with someone who is deceased. It's not an occurrence that happens every so often, this is every weekend this is happening.[24]

## Risk Management

It's clear that alcohol is an active co-conspirator with a young brain, trying to reconcile urges and impulses with self-control (see p. 92). But how and why do these urges and impulses arrive and how and when does sufficient self-control develop to compensate for this?

Those who've had the luxury of not noticing adolescent changes in behaviour – at least not yet – can appreciate at a distance that when adolescence arrives, there is a changing balance between competing brain systems: those involved in amplifying emotion, versus those involved in regulating strong emotion.[25] And there is an increase in seeking novelty, taking risks and a decided swing towards peer-based interactions.

These behaviours are designed to encourage separating from the comfort and safety of the family to explore new environments and seek unrelated mates. However, at the same time, these potentially adaptive behaviours also pose substantial dangers, especially when mixed with modern temptations, including easy access to drugs, firearms, high-speed cars, boats and motorcycles and, of course, alcohol.

Adolescence is a teetering bridge – a time of great risk and great opportunity. The risk trajectory follows a rapid rise in early adolescence, with a peak in late adolescence and early adulthood. And what has traditionally been viewed as a parenting and social issue has now also become of great interest to the field of developmental psychobiology.

From an evolutionary point of view, it's not surprising that the brain is particularly changeable during adolescence – a time when children need to learn how to survive independently in whatever environment they find themselves. The balance among frontal (executive-control) and limbic (emotional) systems alters. Unfortunately for parents and teachers, however, one of the last parts of the brain to complete the maturation process is the prefrontal cortex – the part of the brain that is responsible for planning, judgment ... and self-control. So while adolescents are capable of experiencing very strong and complex emotions and passions, their prefrontal cortex hasn't caught up with them yet and it's as though they don't have the brakes to slow those emotions down. This helps to explain the often irrational behaviour of teenagers – the mood swings, as well as the risks they're often too willing to take.

More recently, in 2010, scientists separated out the timelines for the rise in impulsive behaviour and 'reward seeking' – in other words, looking for kicks. These two have different neurological underpinnings, and the difference in their timetables helps to account for heightened risk taking during adolescence. In fact, a study at Temple University in Philadelphia which examined age differences in reward seeking and impulsivity in 935 people between the ages of ten and thirty found that age differences in reward seeking followed a curvilinear pattern, increasing between pre- and mid-adolescence and declining thereafter. In contrast, age differences in impulsivity followed a linear pattern, with impulsivity declining steadily from age ten on. So heightened vulnerability to risk taking in middle adolescence may be due to the combination of a relatively higher desire to seek rewards and a still-maturing capacity for self-control.[26]

With a rough idea of both the timelines and the biological processes that make them happen, some parents might now assume they can make a note in their forward planners, bury their heads in the sand and just enjoy the calm years until teenage birthday(s) arrive. Unfortunately though, yet another group of scientists are finding that the unscheduled early arrival of puberty is putting paid to all of our best-laid plans.

Early puberty is 'uniquely associated with substantial risk'. There is considerable evidence that early onset of puberty has effects on psychological wellbeing and problem behaviours 'during the adolescent decade and possibly beyond'. In fact, researchers believe that 'off-time' sexual development, that is being earlier or later than one's peers, 'confers risk for serious problems ... and disorders'.[27]

A study by the British government in 2010 is a variation on a commonly observed theme. The relationship between risk taking and alcohol is a two-way affair. They found that young people who engaged in other risky behaviours were more likely to try alcohol, and that trying alcohol was 'strongly predictive of engaging in other risky behaviours, especially criminal behaviour'.

Among young people who had not previously tried alcohol at ages fourteen or fifteen, playing truant, shoplifting, going to parties or pubs and hanging around near home or in town, smoking and trying cannabis 'were all predictive of trying alcohol in the following year ... The relationship between trying alcohol and the subsequent engagement in other risky behaviours tended to be stronger than the other way around, with trying alcohol strongly predictive of increased truancy, smoking, trying cannabis and, particularly, criminal behaviour.'[28]

And so it's very clear that alcohol increases our children's risk of accidents and their vulnerability to crime. But aside from changing our views and laws regarding children's access and sense of entitlement to alcohol, there are bigger issues.

Boys, in particular, must have their risk channelled into less overtly destructive forms. Of course, this leads to a much wider question about young people being able to express risk differently. Children and young people have (justifiably) come to expect stimulation and entertainment to be available and provided most of their waking lives from outside sources, including drinking and going to pubs and clubs. And acquiring a buzz from alternative healthier sources would involve a complete overhaul of our values, economy and lifestyles. Still, with alcohol as the elephant in the room at most crime scenes, it's time we started prosecuting this accomplice who's managed to keep a low profile for far too long.

judgment: 'Children and their parents and carers are advised that an alcohol-free childhood is the healthiest and best option.'[1]

In a similar vein, the US Surgeon General's Call to Action, published in 2007, also makes strong value judgments: 'Adolescent alcohol use is not an acceptable rite of passage but a serious threat to adolescent development and health ... Underage alcohol use is not inevitable, and schools, parents, and other adults are not powerless to stop it.'[2]

As parents and health educators, we must stop being duped into believing that appearing morally passive and value-free is the best way to connect with children – because it's not. And we have to remind ourselves that our children should not be treated as adults – because they're not.

It's important to decide what you think is an acceptable stance on alcohol to adopt for your child. It's important to have clear boundaries: even if children rebel, they need something to rebel against. Alcohol is one area we should feel entitled to enjoy as adults, yet not condone for our children if we don't feel comfortable about them drinking. It is not in the slightest bit hypocritical, for the medical reasons outlined in this book, although it is, of course, always important to set a good example when we drink.

Our children have to learn the protocols of life to become well-balanced adults. And that means learning that sometimes other people can have or do things that they may not. Our children must understand that different people have different values and rules, and that attitude to alcohol is merely one of these.

When I was speaking to a group of parents recently about children and alcohol, one said he let his adolescent sons have some wine while he was drinking in order to 'demystify' alcohol. Another father said his ten- and twelve-year-old sons wanted sips of his wine, so he allowed it 'in order to satisfy their curiosity'. However, this logic could be applied to all manner of adult things – smoking a Marlboro cigarette, for example – and, as such, it is flawed. Such approaches are indicative of parents shying away from being authoritative and simply saying, 'No. I'm drinking my wine because I'm an adult.'

Ultimately, after considering all of the studies on parenting style and underage drinking in Chapter 3, is there anything wrong with a more traditional situation, whereby we know our child *will* experiment with alcohol behind the bike shed, *they* know we know what they're doing, but they *also* know that we don't condone it? And this point is vital because our values will permeate their subconscious.

Drinking is bad for the health of children and young people. And the longer our children delay alcohol use, the less likely they are to develop any problems associated with it. Parents can have a major impact on their children's drinking, especially during the pre- and early teen years. And you have more influence on your child's values and decisions about drinking before he or she begins to use alcohol than after.

## Pocket Money

Some aspects of parenting with regard to alcohol are very pedestrian. In a study in 2007 – 'Predictors of risky alcohol consumption in schoolchildren and their implications for preventing alcohol-related harm' – researchers discovered that when it comes to preventing alcohol misuse, cash is king. They studied fifteen- to sixteen-year-old drinkers in England and of almost 90 per cent who admitted to drinking alcohol, 38 per cent binged, 24 per cent drank frequently and 50 per cent drank in public. Around a third bought their own alcohol, making them six times more likely to drink in public and more than twice as likely to binge and drink frequently as those who'd had alcohol bought for them.

Alcohol misuse was linked with having more than £10 per week spending money, and interventions to 'reduce money available to young people' or to advise parents on improving their monitoring of what adolescents spend money on were strongly recommended. Teenagers (aged twelve to sixteen) in the UK typically received almost £10 pocket money a week from parents, while over a third (37 per cent) of fourteen- to fifteen-year-olds worked in a regular paid job during school term time.[3]

## Talking the Talk

Now that the government wants us to talk to our children about alcohol, we need a briefing document. And here it is, making this chapter (deliberately) different from the others. It is not intended as an instruction manual, and it may sound slightly corny at times, but it should provide food for thought and covers the main areas that need to be considered. Connecting with our children is hardly a fine science, and I have drawn heavily on organisations, especially the US National Institutes of Health, to help me outline some of the principles and factors involved in fulfilling the government's brief.[4]

Early adolescence is a time of enormous and often confusing changes for your child, which makes it a challenging time for them and also for you. But being tuned in to what it's like to be an adolescent can help you stay closer to your child and have more influence on the choices he or she makes – including decisions about using alcohol.

Your child looks to you for guidance and support in making life decisions – including those about alcohol. 'But my child isn't drinking yet,' you may say. 'Isn't it a little early to be concerned about drinking?' Absolutely not! Some children begin experimenting with alcohol at a very young age – and even if your child is not drinking yet, they may be receiving pressure from others to do so. Act now. Keeping quiet about how you feel about this issue may give them the impression that you think alcohol use is OK for children.

As children approach adolescence, friends exert a lot of influence. Fitting in is a chief priority for them, and parents often feel displaced. However, children will listen. Study after study shows that even during the late teenage years, parents have an enormous impact on their children's behaviour.

The bottom line is that many young teenagers (meaning thirteen- to fourteen-year-olds) don't yet drink. And parents' disapproval of underage alcohol use is the key reason for them making this choice. So make no mistake: you *can* make a difference.

Trust your instincts. Choose ideas you are comfortable with, and use your own style in carrying out the approaches you find useful. As with talking to your child about sex and 'where babies come from', you can adjust the level of sophistication and detail according to their age, experience, degree of development and maturity and your values.

## Physical changes

Most ten- to fourteen-year-olds experience rapid increases in height and weight as well as the beginnings of sexual development. As a result, many feel more self-conscious about their bodies than they did when they were younger and begin to question whether they are 'good enough' – that is tall enough, slender enough, strong enough, attractive enough – compared with their peers. And a young teenager who feels he or she doesn't measure up in some way is more likely to do things to try to please friends, including experimenting with alcohol.

During this vulnerable time, it is particularly important to let your children know that in your eyes, they do measure up – and that you care about them deeply.

## Thinking skills

Most young teenagers are still very 'now' oriented, and are only just beginning to understand that their actions – such as drinking – have consequences. They also tend to believe that bad things won't happen to them, which helps to explain why they often take risks. Therefore, it is very important for adults to invest time in helping children understand how and why alcohol-related risks do apply to them.

## Social and emotional changes

Young teenagers increasingly look to friends and the media for clues on how to behave and begin to question adults' values and rules. It's not surprising then that parents often experience conflict with their children as they go through early adolescence. During this potentially turbulent period, perhaps your most difficult – and sometimes exasperating – challenge is to try to respect your child's growing drive for independence, while still providing support and limits.

## Bonding

The best way to influence your child to avoid drinking is to have a strong, trusting relationship with them. Research shows that teens are much more likely to delay drinking when they feel they have a close, supportive tie with a parent or guardian. Moreover, if your son or daughter eventually does begin to drink, a good relationship with you will help protect him or her from developing alcohol-related problems.

And the opposite is also true: when the relationship between a parent and teenager is full of conflict or is very distant, the teenager is more likely to use alcohol and to develop drinking-related problems. This connection between the parent–child relationship and a child's drinking habits makes a lot of sense when you think about it. Firstly, when children have a strong bond with a parent, they are apt to feel better about themselves and therefore be less likely to give in to peer pressure. Secondly, a good relationship with you is likely to influence your children to try to live up to your expectations because they want to maintain their close tie with you and gain that all-important parental approval rating.

In short, show you care. Even though young teenagers may not always show it, they still need to know they are important to their parents. Make it a point to regularly spend one-on-one time with your child when you can provide them with loving, undivided attention – say, taking a walk or a bike ride together, having a quiet dinner out or even a biscuit-baking session.

Try to make it easy for your adolescent to talk honestly with you. Developing open, trusting communication between the two of you is essential in helping your child to avoid and delay drinking. If they feel comfortable talking openly with you, you'll have a greater chance of guiding him or her towards healthy decision making. The experts emphasise the following points:

- Encourage your child to talk about whatever interests him or her. Listen without interruption and give your child a chance to tell you something new. Your active listening to your child's enthusiasms paves the way for conversations about topics that concern you.
- Ask open-ended questions. Encourage your teen to tell you how he or she thinks and feels about the issue you're

discussing. Avoid questions that have a simple 'Yes' or 'No' answer.

- Control your emotions. If you hear something you don't like, try not to respond immediately with anger. Instead, take a few deep breaths and wait.
- Don't only lecture or try to 'score points' over your teenager by showing how he or she is simply wrong. If you show respect for your child's viewpoint, he or she will be more likely to listen to and respect yours.
- Draw the line. Set clear, realistic expectations for your child's behaviour. Establish suitable consequences for breaking rules and try to enforce them consistently.
- Offer acceptance. Make sure your teenager knows that you appreciate their efforts, as well as their accomplishments.
- Adjust the apron strings or lead. Understand that your child is growing up. This doesn't mean a hands-off attitude, but as you guide your child's behaviour, also make an effort to respect his or her growing need for independence and privacy.

## Talking to Your Adolescent About Alcohol

Bringing up the subject of alcohol isn't always easy. Your young teenager may try to avoid the discussion, and you may feel unsure about how to proceed. To improve your chances for a useful conversation, bear the following in mind:

- Take some time to think through the issues you want to discuss before you start talking.
- Think about how your child might react to what you're going to say and ways you might respond to their questions and feelings.
- Choose a time to talk when both you and your child have some time and are feeling relaxed. Remember, you don't need to cover everything at once. In fact, you're likely to have a greater impact on your child's drinking by having a number of talks about alcohol use throughout his or her adolescence. Think of this opening discussion with your child as the first part of an ongoing conversation, not just a lecture.
- Be sure to state clearly your own expectations regarding your child's drinking and to establish consequences for breaking rules. Your values and attitudes count with them, even though they may not always – or ever – show it.

## Your child's views

Ask your young teenager what they know about alcohol and what they think about teenage drinking. Ask them why they think children drink. Listen carefully without interrupting. Not only will this approach help your child to feel heard and respected, but it can serve as a natural 'lead-in' to discussing alcohol topics.

## Important facts about alcohol

Although many teenagers believe they already know everything about alcohol, myths and misinformation abound. This book offers a veritable menu of information you can share with your child; you might even leave them with a copy, pointing out the areas they should skim. Often, it's the shallow effects of alcohol that concern teenagers. So while liver disease might not scare them, alcohol causing body fat will.

## Inoculation Against the Media

The media's glamorous portrayal of alcohol encourages many teenagers to believe that drinking will make them popular, attractive, happy and cool. As Chapter 2 made clear, research shows that adolescents who expect such positive effects are more likely to drink at early ages. However, you can help to combat these dangerous myths by watching TV shows and movie videos with your child and discussing/counteracting how alcohol is portrayed in them.

## *Good* Reasons Not to Drink

In talking to your child about reasons to avoid alcohol, don't rely on unrealistic scare tactics. Most young teenagers are aware that many people drink, apparently without overt problems. And because many of the effects of alcohol use are subtle or insidious, often appearing only years later, your child is unlikely to associate them with drinking, wrongly assuming that any significant problem would be visible, here and now. So it is important to discuss the consequences of alcohol use without overstating the case.

## Family history

If there is a history of alcoholism in your family, you may want to tell your child that they are more at risk of developing a drinking problem because of this; your child needs to know that for him or her, drinking may carry special risks.

## Their developing brain

Children may absorb the concept that alcohol affects young people very differently from adults and that this is the basic reason why you want them to delay their drinking (or even avoid it altogether). For example, in Chapter 4, it is made very clear that drinking while the brain is still maturing may lead to long-lasting intellectual consequences and may even increase the likelihood of developing alcohol dependence later in life.

I don't know of any parents in England who encourage their children to consider being a non-drinker when they grow up. We bow to what we see as the inevitable. But while children are developing, absorbing our values and ideas and looking at various options, we should include as one of those options the possibility of them being a non-drinker – where alcohol may be seen as fun, but not as necessary to have fun.

### Appeal to their self-respect . . .

Letting your child know that they are too smart and have too much going for them to need the crutch of alcohol is a good approach.

### . . . and their self-image

Teens also pay attention to ways in which alcohol might cause them to do something embarrassing that might damage their self-respect and important relationships.

I enjoy telling boys that girls do not find binge drinkers as sexually attractive as non-binge drinkers by pointing out that I have yet to see a lonely-hearts ad stating: 'Large-breasted, beautiful woman seeks tall, dark, legless binge drinker.' I also mention that boxers and cage fighters tend not to drink, and certainly not within a few days of a fight. 'If you want to win a fight,' I tell them, 'stay sober and let the other guy get drunk, overly confident and slow. Then you can beat him up far more easily!' Of course, I then immediately advocate peace and love wherever possible.

## Mum, Dad, Did You Drink When You Were Kids?

This is the question many parents dread – yet it is highly likely to come up in any family discussion of alcohol. The reality is that many of us parents did drink underage. So how can we be honest with a child without sounding hypocritical?

This is a judgment call. If you believe that your drinking history should not be part of the discussion, you can simply tell your child that you choose not to share it. Another approach is to admit to them that you did do some drinking as a teenager, but that it was a mistake, and give them an example of an embarrassing or painful moment that occurred because of your drinking. This line may help your child better understand that youthful alcohol use really does have negative consequences.

But perhaps most important is the plain fact that we simply didn't know then that alcohol has the profound effects that it does. In the same way that we didn't realise the effects of many other things, such as smoking, passive smoke, sun tanning, about which we have since changed our views.

## Preparing for Peer Pressure

The off-the-shelf phrase 'peer pressure' does have to be considered here, despite the fact that it's abstract and outside of our direct control. It's not enough to tell a young teenager that he or she should avoid alcohol – we also need to help them figure out how.

What should your daughter say when she goes to a party and a friend offers her a vodka and Red Bull? What should your son do if he finds himself in someone's home where kids are passing around a bottle of wine and the parents are nowhere in sight? And what should their response be if they are offered a ride home with an older friend who has been drinking?

Brainstorming with your child about ways in which they might handle these and other difficult situations shows that you are willing to support them. Tell them, for example, 'If you find yourself at a home where kids are drinking, call me and I'll pick you up – and there will be no scolding or punishment.' The more prepared your child is, the better they will be at handling high-pressure situations that involve drinking. Discussing possible scenarios ahead of time will help them to cultivate a nimbleness and agility in negotiating the world of underage drinkers.

At some point, your child *will* be offered alcohol. Teenagers say they prefer quick 'one-liners' that allow them to pass up a drink without making a big scene. It will probably work best if your teenager takes the lead in thinking up their own comebacks to offers of drinks, so that they feel comfortable using them. But to get the brainstorming started, here are some simple pressure-busters – from the mildest to the most assertive:

- No thanks.
- I don't feel like it – do you have any Coke (that's cola not -caine)?
- Alcohol's not my thing.
- Are you talking to me? FORGET it!
- Why do you *keep* pressuring me when I've said NO?
- Back [piss] off!

At a more general level, this discussion is about your child developing a natural assertiveness to protect their own values and self. They can learn to resist alcohol, or indeed anything else they may feel pressured into. We have to let them know that the best way to say 'No' is to be assertive – that is, say 'No' and mean it. Stand up straight. Make eye contact. Say how you feel. Don't make excuses. Stand up for yourself.

## Parent Plan

While parent–child conversations about drinking are essential, talking alone isn't enough – we also need to take concrete action to help our children resist alcohol. Research strongly shows that active, supportive involvement by parents and other key adult figures can help adolescents avoid underage drinking and prevent later alcohol misuse. US national surveys have found that between 64 and 71 per cent of thirteen-year-olds say alcohol is 'fairly easy' or 'very easy' to get hold of. We know that in Britain it's far easier, and in Switzerland, for example, the legal drinking age is sixteen.

The message is clear: young teenagers still need a great deal of adult supervision. Here are some ways to provide it:

### Monitor alcohol use in the home

If you keep alcohol in your home, keep track of the supply. Make clear to your child that you don't allow unchaperoned parties or other teenage gatherings in your home. If possible, however, you should encourage them to invite friends over when you're at home. The more entertaining your child does at home, the more you'll know about their friends and activities.

### Connect with other parents

Getting to know other parents and guardians can help you to keep closer tabs on your child. Friendly relations can make it easier for you to call the parent of a teenager who is having a party to ask if a responsible adult will be in the house and ensure that alcohol will not be available. You're likely to find out that you're not the only adult who wants to prevent underage drinking – many other parents are beginning to share this concern.

### Keep track of your child's activities

Be aware of your adolescent's plans and whereabouts. Generally, your child will be more open to supervision if they feel you're keeping tabs because you care, not because you distrust them.

### Establish family rules about underage drinking

The studies make it clear that when parents establish clear 'no alcohol' rules and expectations, their children are less likely to begin drinking. While each family should formulate agreements about teenage drinking that reflect their own beliefs and values, here are some ideas:

- Children are not to drink until they are a certain age.
- Older siblings will not encourage younger brothers or sisters to drink and will not give them alcohol.
- Children will not ride in a car with a driver who has been drinking.

Once you've chosen rules that work for your family, you will need to establish some fitting consequences for breaking them, with penalties that you are willing to carry out. Be careful though not to make the consequences so harsh that they become a barrier to open communication between you and your child. The idea is to make the penalty 'sting' just enough to make your child think twice about breaking the rule. A possible consequence might be temporary restrictions on their socialising, for example. And you must be prepared also to enforce the consequences you've established consistently. If your child knows that they will lose certain privileges each and every time an alcohol-use rule is broken, they will be more likely to keep it.

## Set a good example

Parents and guardians are important role models for their children – even those who are fast becoming teenagers. Studies indicate that if a parent uses alcohol, his or her children are more likely to drink themselves. But even if we drink, which most parents do, there may be ways to lessen the likelihood that our child will do so, or do so harmfully. Here are some suggestions:

- Drink in moderation – at least when your children are around, you should try to keep a low drinking profile, in the same way as most of us keep other aspects of our adult lives private.
- Don't communicate to your child that alcohol is a good way to handle problems. For example, don't come home from work and say, 'I had a rotten day. I need a drink.' (Or the more English, 'I could *murder* a pint!') Let your child see that you have other, healthier ways to cope with stress, such as exercise, listening to music, or talking things over with a spouse, partner or friend.
- Don't tell your children stories about your own drinking in a way that conveys the message that alcohol use is funny or glamorous. The 'Remember-that-stupid–thing-I-did-last-year-when-I-was-pissed'-type stories should be for adult ears only.
- Never drink and drive or ride in a car with a driver who has been drinking.
- When entertaining other adults, make available alcohol-free drinks and plenty of food. If anyone drinks too much at your party, make arrangements for them to get home safely.

## Don't support underage drinking

Your own attitude and behaviour towards underage drinking can influence your child, including showing acceptance of it. In addition, never serve alcohol to your child's underage friends. Research shows that kids whose parents or friends' parents provide alcohol for teenage get-togethers are more likely to engage in heavier drinking, to drink more often and to be involved in traffic accidents.

## The school curriculum

It's worthwhile joining school and community efforts to discourage drinking by teenagers. By working with schools and other members of the community, you can help to develop policies to reduce alcohol availability to teenagers and to enforce consequences for underage drinking.

Personal Social and Health Education (PSHE) is part of our children's school curriculum. Ask about this and try to ensure that they are learning about alcohol early and often, ideally starting in primary school. Do not wait until they are already exposed to underage drinkers.

## Help your child to build healthy friendships

If your child's friends drink, your child is more likely to drink too. So it makes sense to try to encourage your young teenager to develop friendships with children who do not drink and who are otherwise healthy influences.

A good first step is simply to get to know your child's friends better. You can then invite those you approve of to family get-togethers and outings and find other ways to encourage your child to spend time with them.

Also, you can talk directly with your child about the qualities in a friend that really count, such as trustworthiness and kindness, rather than mere popularity or coolness ... but often they have to learn this the hard way.

When you disapprove of one of your child's friends, the

situation can be more difficult to handle. While it may be tempting to forbid your child to see a particular child, such a move may make them even more determined to hang out with this friend. Instead, try pointing out your reservations in a caring, supportive way. You can also limit your child's time with that friend, say through your family rules, such as how after-school time may be spent or how late your child may stay out in the evening.

## Encourage healthy alternatives to alcohol

One reason why children drink is to beat boredom. Therefore, it makes sense to encourage your child to participate in supervised after-school and weekend activities that are challenging and fun. According to a recent survey of pre-teens, the availability of enjoyable activities is a big reason for them deciding not to use alcohol.

If your community doesn't offer many supervised activities, consider getting together with other parents and young teenagers to help create some. Start by asking your child and their friends what they would like to do; they will be more likely to take part in activities that truly interest them. Find out whether your school, community organisation, or church can help you by sponsoring a project.

Sport and physical activity seem to prevent depression (see p. 157) and raise self-esteem. And the social interaction and camaraderie integral to team sports can build on these positive effects. Also, despite the sometimes negative image associated with pop and rock bands, the discipline and goals of being in a band are healthy and can help to keep children focused on constructive activities.

### Party parents

The US National Institutes of Health kindly offers advice on 'How to Host a Teen Party':

- Agree on a guest list – and don't admit party crashers.
- Discuss ground rules with your child before the party.
- Encourage your teen to plan the party with a responsible friend so that they will have support if problems arise.
- Brainstorm fun activities for the party.
- If a guest brings alcohol into your house, ask them to leave.
- Serve plenty of snacks and non-alcoholic drinks.
- Be visible and available – but don't join the party.

## University

As mentioned in Chapter 3, the children of parents who disapprove completely of underage alcohol use tend to engage in less drinking and less binge drinking at university. And conversely, parents' permissiveness towards teenage drinking is a significant risk factor for later binge drinking.[5]

Anecdotal evidence suggests that the early weeks of the autumn term are critical to a first-year student's academic success, and because many students begin heavy drinking during these early days, the potential is there for excessive alcohol use to interfere with this, as well as with successful adaptation to campus life. The transition to university is often difficult (in the US, about one-third of first-year students fail to enrol for their second year) and the autumn term is the ideal time for parents to revisit discussions about university drinking, reminding their children of the consequences of too much alcohol (date rape, violence and academic failure, to name a few).

Some first-year students who live on campus may be at particular risk for alcohol misuse. During their secondary-school years, those who go on to university tend to drink less than their non-university-bound classmates. However, during subsequent years, the heavy drinking rates of uni students surpass those of their non-uni peers. The rapid increase in heavy drinking over a relatively short period of time can contribute to serious difficulties with the transition to university and lay the groundwork for future problems.

During the crucial early university weeks, parents can do a variety of things to stay involved, such as enquiring about campus alcohol policies, calling their child frequently and asking about roommates and living arrangements.

All we can do in a booze-adoring culture is reduce the risk that our children won't misuse alcohol, but with the greatest will in the world we cannot prevent it with certainty. So we need to be equipped to deal with problems if they arise, and this is what will be covered in the next chapter.

Ten

# Plan B: Dealing with Alcohol Problems

Few adults reading this would not have experimented with alcohol when they were underage. But experiments can go wrong or become a regular hobby. As discussed in Chapter 3, our children's genes predispose how they will take to alcohol, how much they can 'handle' before being drunk and how easily they can become addicted, now and in future. And while the main point of this book is to prevent problems altogether, by reducing our children's contact with alcohol until they are older, it's vital to notice when one is developing so that it is less likely to become a long-term adult issue.

It is common among adolescents to experiment with alcohol. But while many of them experiment and stop or use occasionally with no long-term harmful effects, there are some who develop a dependency upon alcohol, leading to significant problems down the road. Knowing some of the signs of teenage alcohol abuse can help parents to realise there's an issue early so that they can work to prevent it from developing further. Although parents have different policies when it comes to adolescent drinking, from a strict zero-tolerance approach, to promoting moderation, problem drinking should always be dealt with quickly and without compromise.

Problems with alcohol have been 'medicalised' or 'psychotherapised'. Doctors need, understandably, to analyse, categorise and refine the diagnostic criteria for specific alcohol problems; this doesn't mean, however, that they don't consider or appreciate the emotional reasons behind them as well. It is simply a method that allows them to help people to the best of their ability. Therapists and former alcoholics tend to focus on the emotional road that leads to drinking and the emotional healing road that brings someone back. Arguments continue about exactly what constitutes being an 'alcoholic' and there are also some complaints about the use of such a term.

If we feel our child (or, in the case of a teacher, their student) appears to have an alcohol problem, that is often good enough grounds for concern. Alcohol problems can be the result of unhappiness or of predisposing genes enjoying too much of a good time (see p. 80) with a substance that can be highly addictive at too young an age. Or it can be both, in various measures. And peer influence plays a big role too, interacting with the other two factors.

While we needn't be overly concerned with categories, it is worth taking a brief look at how the professionals regard them, as this will give you a better feel for the problem. Alcohol issues are increasingly referred to as alcohol-use disorders (AUDs), broadly divided into alcohol abuse ('hazardous'/'harmful' drinking) and 'alcohol dependence'. (These categories are discussed in more detail below.) The diagnostic criteria for AUDs have largely been developed based on research and clinical experience with adults.

The World Health Organization defines these drinking patterns thus:

- *Hazardous drinking* is a pattern of alcohol consumption that increases the risk of harmful consequences for the user or others. Hazardous-drinking patterns are of public health significance despite the absence of any current disorder in the individual user.
- *Harmful use* refers to alcohol consumption that has consequences for physical and mental health. Some would also consider social consequences among the harms caused by alcohol.
- *Alcohol dependence* is a cluster of behavioural, cognitive and physiological phenomena that may develop after repeated alcohol use. Typically, these phenomena include a strong desire to consume alcohol, impaired control over its use, persistent drinking despite harmful consequences, a higher priority given to drinking than to other activities and obligations, increased alcohol tolerance and a physical withdrawal reaction when alcohol use is discontinued.

The American Psychiatric Association divides AUDs into two primary types: alcohol abuse and alcohol dependence. A person receives a diagnosis of alcohol abuse if he or she experiences at least one of four abuse symptoms (i.e. role impairment, hazardous use, legal problems, and social problems) that lead to 'clinically significant impairment or distress'.

For a diagnosis of alcohol dependence, a person must exhibit within a twelve-month period at least three of the following seven dependence symptoms:

- Tolerance
- Withdrawal or drinking to avoid or relieve withdrawal
- Drinking larger amounts or for a longer period than intended
- Unsuccessful attempts or a repeated desire to quit or to cut down on drinking
- Much time spent using alcohol
- Reduced social or recreational activities in favour of alcohol use
- Continued alcohol use despite psychological or physical problems.[1]

The National Institute for Health and Clinical Excellence (NICE) in Britain views 'alcohol misuse' along similar lines, but acknowledges that there is overlap between the categories of AUDs: 'Although alcohol dependence is defined in categorical terms for diagnostic and statistical purposes as being either present or absent, in reality dependence exists on a continuum of severity.'[2]

People may drink in a way which is hazardous, abusive, harmful and they can also be dependent on alcohol. And because these classifications were developed primarily from research on adults, many adolescents do not fit neatly into them or are not deemed to exhibit enough of the right symptoms to warrant a diagnostic category. In fact, some researchers coined the term 'diagnostic orphans' to describe adolescents with one or two of the alcohol-dependence symptoms and none of the alcohol-abuse symptoms, who therefore do not qualify for either diagnosis.[3]

In general terms, children who are at the highest risk for alcohol-related problems are those who have:

- begun using alcohol before the age of fifteen
- a parent who is a problem drinker or an alcoholic
- close friends who use alcohol and/or other drugs
- been aggressive, antisocial or hard to control from an early age
- experienced childhood abuse and/or other major traumas
- current behavioural problems and/or are failing at school
- parents who do not support them, communicate openly with them or keep track of their behaviour or whereabouts
- experience of ongoing hostility or rejection from parents and/or overly harsh, inconsistent discipline.

The more of the above criteria that apply to a child, the greater the chances are that they will develop problems with alcohol. Of course, having one or more risk factors does not mean that your child definitely will develop a drinking problem, but it does suggest that you may need to act now to help protect your child from later problems.

## Signs and Symptoms: How to Tell if Your Child May Be in Trouble with Alcohol

Just as categorising alcohol problems is not a fine science, neither is the idea that there are any clear signs and symptoms that your child has a problem with alcohol. Although the following signs (compiled by the National Council on Alcoholism and Drug Dependence) may indicate a problem, some also reflect normal teenage 'growing pains'. However, it's generally thought that a drinking problem is more likely to be indicated if you notice several of these signs at the same time, if they occur suddenly and if some are extreme in nature:

- Smell of alcohol on breath, or sudden, frequent use of breath mints.
- Abrupt changes in mood or attitude.
- Sudden decline in attendance or performance at school.
- Losing interest in school, sports or other activities that used to be important.
- Sudden resistance to discipline at school.
- Uncharacteristic withdrawal from family, friends or interests.
- Heightened secrecy about actions or possessions.
- Associating with a new group of friends whom your child refuses to discuss.

Aside from the above, parents should also use their instincts as to whether something simply seems wrong.

Obviously, if you can talk to your child and find out about their drinking in a straightforward way, so much the better. But secrecy is often the norm, and, as is the case with adults who may be 'functional alcoholics' but are in denial, adolescents may find it even more incomprehensible still to see themselves as 'alcoholic'. Talking to your child from a point of concern and desire to understand their experience, the pressures in their life and the circumstances they find themselves in may lead them to drop their defences. As will alluding to the fact that you may have gone through similar phases yourself as a teenager and may have faced similar dilemmas.

Some of the key themes that they may recognise is drinking to change the way you feel – not in the pleasurable sense, but to blot out pain, anxiety or unhappiness. Is their drinking having an impact on other areas of their life: social, school, romantic, physical? It may sound corny, but there are self-tests your adolescent can use to see for themselves whether they have an alcohol problem (for example, the National Council on Alcoholism and Drug Dependence).[4]

## How Are Alcohol and Drugs Affecting Your Life? A Self-Test for Teenagers

1 Do you use alcohol or other drugs to build self-confidence?
2 Do you ever drink or get high immediately after you have a problem at home or at school?

3 Have you ever missed school due to alcohol or other drugs?
4 Does it bother you if someone says that you use too much alcohol or other drugs?
5 Have you started hanging out with a heavy-drinking or drug-using crowd?
6 Is alcohol or other drugs affecting your reputation?
7 Do you feel guilty or a bit low after using alcohol or other drugs?
8 Do you feel more at ease on a date when drinking or using other drugs?
9 Have you got into trouble at home for using alcohol or other drugs?
10 Do you borrow money or 'do without' other things to buy alcohol and other drugs?
11 Do you feel a sense of power when you use alcohol or other drugs?
12 Have you lost friends since you started using alcohol or other drugs?
13 Do your friends use less alcohol or other drugs than you do?
14 Do you drink or use other drugs until your supply is all gone?
15 Do you ever wake up and wonder what happened the night before?
16 Have you ever been arrested or hospitalised due to alcohol or use of illicit drugs?

17 Do you 'turn off' any studies or lectures about alcohol or illicit drug use?
18 Do you think you have a problem with alcohol or other drugs?
19 Has there ever been someone in your family with a drinking or other drug problem?
20 Could you have a problem with alcohol or other drugs?

## Scoring

If you answered 'Yes' to any three of the above questions, you may be at risk for developing alcoholism and/or dependence on another drug. If you answer 'Yes' to five of these questions, you should seek professional help immediately.

An even simpler questionnaire is as follows:

1 How many drinks does it take before you begin to feel the first effects of alcohol? 1 2 3 4
2 Have your friends or relatives worried about your drinking in the past year? Yes/No
3 Do you sometimes have a drink in the morning when you first get up? Yes/No
4 Are there times when you drink and afterwards can't remember what you said or did? Yes/No
5 Do you sometimes feel the need to cut down on your drinking? Yes/No

### Scoring

1 Three or more drinks = 2 points

2 Yes = 1 point

3 Yes = 1 point

4 Yes = 1 point

5 Yes = 1 point

If you score 3 or more points, or if your answer to any one of the questions concerns you, you may be using alcohol in ways that are harmful and should try to get some help.[5]

## Seeking Help

Of the million people aged between sixteen and sixty-five who are alcohol dependent in England, only about 6 per cent per year receive treatment. Reasons for this include the often long period between developing alcohol dependence and seeking help, and the limited availability of specialist alcohol treatment services in some parts of the country. Additionally, alcohol misuse is under-identified by health and social-care professionals, leading to missed opportunities to provide effective interventions.[6]

Even though we may have house rules on alcohol and consequences for breaking our house rules, we need to reconcile any formal position we may take regarding our children drinking in general terms with the current reality that many are now in trouble with alcohol. They need our help but they need to realise this, and feel comfortable asking for it or allowing us in to do it. Again, this is the human condition – or worse, the adolescent condition – and so, again, communicating about the delicate issue of a possible alcohol problem is not a fine science.

On a practical level, if you believe your child has an alcohol problem, you need to remember that it is an addictive drug. According to how dependent they are on alcohol, you should think about not making it too accessible in your home for the time being. A fully stocked drinks cabinet is not going to be helpful. Explaining the problem to brothers and sisters and ensuring that they don't inadvertently wave alcohol under their sibling's nose is also a good idea. Teenagers may see it as a joke and not take it seriously enough if it is not made clear to them.

It can be almost impossible for us, as parents, to be objective about our own children, being, as we are, emotionally involved and invested in them. Therefore, we need help and support from people, professionals and organisations, who are familiar with adolescent alcoholism. This is especially so when there is only one parent, particularly if that lone parent is a mother and the problem drinker is a boy. A big, strong, angry teenage boy who doesn't drink can be difficult enough to deal with, but for a lone mother, trying to cope with the same boy who *is* drinking can be physically unmanageable. (There is a list of organisations and sources of help for *all* parents and for children in the Resources section of this book – see p. 262.) Parents also mustn't overlook the need to know that they are not the only ones whose child has a drinking problem; comparing notes with other parents about how they've coped and dealt with the problem can be very helpful.

Feeling guilty or at fault is often entirely unjustified and only adds to the worry and anguish over the suspicion or fact that a child has an alcohol problem. But while our therapy culture may have made our society more emotionally literate in some ways, it has also caused a reading for deeper meaning behind a problem that might actually be shallower than we've been led to believe. As discussed in Chapter 7, a child may indeed drink to change the way they feel, to blot out the unhappiness about a home situation. But they may also drink purely for pleasure, to get high because it simply feels good. After all, no deep analysis is required to understand a nicotine addiction.

These are some of the avenues open to you when it comes to finding professional help:

- **A dedicated alcohol advisory service.** It is possible, in the first instance, to go directly to a service which can help without going through your GP (if, say, you are concerned that there will be an official government record of your child's alcohol problem). For example, in the UK, Alcohol Concern provides an online Alcohol Services Directory covering a wide range of services from residential rehabilitation to telephone counselling, available for problem drinkers and their families. The services include those across the voluntary, health and private sectors in England and Wales. You can also contact your local health authority for alcohol services in your area.

- **Your local pharmacy.** Ask at the counter for advice on services and where you can access help.

Contact details for services that may be useful are provided in the Resources section on pp. 262–8. If you call them, they will be able to advise on what help is available and what your next step should be.

The NICE guidelines, often advising treatment on the NHS, describe some of the important elements in their approach to helping children and young people: 'If the service user agrees, families and carers should have the opportunity to be involved in decisions about treatment and care. For young people under the age of sixteen, parents or guardians should be involved in decisions about treatment and care according to best practice. Families and carers should also be given the information and support they need in their own right.'

In the 'assessment and interventions for children and young people who misuse alcohol', doctors are advised as follows:

- For children and young people aged ten years and older who misuse alcohol offer:
  - individual cognitive behavioural therapy for those with limited additional disorders or illness and good social support
  - multicomponent programmes (such as family therapies) for those with other significant health problems and/or limited social support.

- When there are other problems, illness or disorders which accompany the alcohol misuse doctors are advised as follows:
  - For people who misuse alcohol and have depression or anxiety disorders, treat the alcohol misuse first as this may lead to significant improvement in the depression and anxiety. If depression or anxiety continues after three to four weeks of abstinence from alcohol, undertake an assessment of the depression or anxiety and consider referral and treatment in line with the relevant NICE guideline for the particular disorder.

Whether dealing with alcohol or any other problem, there are different 'brands' of rehab and therapy. And, as with other areas of life, there is rivalry between the various approaches. Some treatments may also include medication which alters the brain's response to alcohol so that drinkers will not derive the same pleasure or 'feel good' factor when they drink, or which helps the brains of those who have become alcohol dependent to work normally again.

Our society is preoccupied with the best 'technique' for everything, including therapy. But many different methods can help and some approaches are more compatible with certain people. This is something that the experts you liaise with will help you to decide. While it's easy for others to reel off platitudes, such as, 'They have to hit rock bottom before they'll stop drinking', as a parent you don't want your child to hit rock bottom in the first place. This may well be the best path for fully grown *adult* recovery, but your child is *not* an adult. And it's important that the effects of alcohol misuse on their developing brain and body are halted during this transition period so that they do not have a lasting effect.

So while it may ultimately come to 'tough love', the first port of call should be love and action.

## Conflict of Interest?

As a general rule, the organisation that provides the public with information about health should not be the same one that manufactures the product that affects it. For example, information about the effects of chocolate and other sweets on diabetes, obesity and cholesterol levels in children should not come from Cadbury, but from independent government scientists and doctors whose research receives no funding whatsoever from the confectionery industry. Yet when it comes to alcohol, this is not the case.

The most high-profile information and advertising children and parents are exposed to comes from Drinkaware.co.uk. The Drinkaware Trust is an independent charitable trust aimed at promoting sensible drinking. You may recognise their tagline from the end of drinks adverts: 'Please drink responsibly'. However, I have yet to meet a parent, teacher or even a doctor who knows that Drinkaware is almost entirely funded by the alcohol industry. In 2010, the House of Commons Select Committee on Health published the fact that the Portman Group (comprising the UK's nine biggest drinks companies) contributes 97 per cent of the money to fund the charity.[1]

It is commendable, of course, that the alcohol industry is promoting 'sensible drinking' and it is not down to them that the Drinkaware Trust occupies the number-one position in alcohol-awareness lessons for the public. Rather, it is the fault of successive governments who have, in the functionary vernacular, 'outsourced' this responsibility, leaving a vital area of health education in the hands of the poachers, as opposed to the gamekeepers.

Take as an example a prominent leaflet in my local GP surgery entitled, 'A Guide for Parents ... [to] help you help your child make the right choices about alcohol – if and when they decide to drink'. Only near the end of the leaflet does it say, in small print: 'The Drinkaware Trust is grateful to Diageo Great Britain for a generous grant towards the development of this leaflet.' Diageo is the world's largest multinational beer, wine and spirits company. The next leaflet patients can choose instead, entitled, 'Discussing Drinking with Your Children', is published by the Portman Group.

The House of Commons Health Committee has started to criticise strongly the current situation. In their 138-page report 'Alcohol' (in 2010), they heard evidence from a range of Britain's top experts and came to some brutal conclusions, which at times make you wince when reading: 'If everyone drank responsibly, the alcohol industry would lose 40 per cent of its sales and some estimates are higher. In formulating its alcohol strategy, the Government must be more sceptical about the industry's claims that it is in favour of responsible drinking.'[2] And, shattering the myth that alcohol is only 'a problem for a small minority, as the drinks industry states', the Committee concludes: 'During this inquiry we heard strikingly contrasting views from Diageo and the Royal College of Physicians.'

Diageo told the House of Commons Health Committee: '... it is wrong to paint Britain as a nation with an alcohol problem', while the Royal College of Physicians said: 'In the UK the health harms caused by alcohol misuse are underestimated and continue to spiral ... Our teenagers have an appalling drink problem.' And the Committee clearly states: 'It is time the Government listened more to the Chief Medical Officer and the President of the Royal College of Physicians and less to the drinks and retail industry.'

In criticising the cosy relationship between government departments and the alcohol industry, the House of Commons Health Committee goes on to brutalise government departments and entire governments in relatively uncompromising tones:

> The alcohol problem in this country reflects a failure of will and competence on the part of government Departments and quangos ... The DCMS [Department for Culture, Media and Sport] has been particularly close to the drinks industry ... Unfortunately, public health has not been a priority for the DCMS ...
>
> The interests of the large pub chains and the promotion of the 'night-time' economy have taken priority; Ofcom, the ASA [Advertsisng Standards Authority] and the Portman Group preside over an advertising and marketing regime which is failing to adequately protect young people. The OFT [Office of Fair Trading] shows a blinkered obsession with competition heedless of concerns about public health. The Treasury for many years pursued a policy of making spirits cheaper in real terms. Collectively Government has failed to address the alcohol problem ...

In Scotland legislation gives licensing authorities the objective of promoting public health.[3]

If we are to protect our children from today's drinking culture effectively, we must heed the strong points made by the House of Commons Health Committee. Our information must in future be completely hygienic and we need to remove and prevent any possible or perceived conflicts of interest that could possibly stand in the way of parents, children, schools, doctors and politicians making *informed* decisions about alcohol and young people. Quite aside from the issue of MPs voting while under the influence when they should be sober and voting on important legislation (see p. 17), they should not be receiving money (either directly or indirectly) or favours from the drinks industry or its various quangos. In short, they shouldn't be having drinks with distillers.

Politicians under the influence of the alcohol industry are also a European problem. The Commissioner of the Directorate for Health and Consumer Protection for the European Commission has said that while the EU was developing a strategy on alcohol, he was 'surprised at the aggressiveness of the lobbying campaign by certain parts of the alcohol industry'; other sources have suggested that this was the strongest lobbying campaign *ever* faced by the Directorate.[4]

Prominent research scientists are starting to complain about the growing involvement of the alcohol industry in important scientific research. Using terms such as 'corporate social responsibility' and 'partnerships with the public health community', the industry funds a variety of scientific activities, such as conferences, research programmes and scientific publications. They even play a role in editing and distributing 'science journals' which are supposed to provide our society's scientists and doctors with 'unbiased studies' about the true effects of alcohol. Examples of secret research by the alcohol industry not released to the public or the scientific community are now being exposed in ethical medical journals. And there are big questions concerning the use of the information gained from this secret research to target vulnerable populations (e.g. adolescents, problem drinkers, women of childbearing age). There is a drive to now sanction researchers who accept money from the alcohol industry and yet keep this fact to themselves when they publish 'objective research' on the effects of alcohol. Yet, we shouldn't be surprised at any of this, remember the same things happening with the tobacco industry?[5]

And here's a sobering thought. As I am writing this, it's just been announced that the Department of Health is putting the alcohol industry at the heart of writing government policy on alcohol. Britain's health secretary has set up an alcohol 'responsibility deal' network with business, co-chaired by ministers, to come up with policies expected to be used in the public health white papers. The alcohol responsibility deal network is chaired by the head of the lobby group the Wine and Spirit Trade Association. Professor Sir Ian Gilmore of the Royal College of Physicians and a leading liver specialist said he was very concerned by the emphasis on 'voluntary partnerships' with industry. A member of the alcohol responsibility deal network,

Professor Gilmore said he had decided to co-operate, but he doubted whether there could be 'a meaningful convergence between the interests of industry and public health, since the priority of the drinks industry was to make money for shareholders while public health demanded a cut in consumption'.[6]

## Inoculation

We must not be left trying to lock the stable door after the horse has bolted with a pint in his hoof. There has to be an urgent shift of attention to making young children aware of the health implications of alcohol long before they reach puberty. And again, parents should not feel as if there is any hypocrisy involved just because they themselves drink at the dinner table. The message for all children and young adults is that we now know that alcohol has especially profound effects on children and young people who are still developing in ways that do not apply to fully grown adults. This is a clear and sensible rationale and a good enough reason to have different policies and approaches to alcohol for children and young adults.

In schools, there has been a great emphasis on drugs awareness, while alcohol has been viewed differently and less urgently. Moreover, it is not even treated as a drug, despite the fact that it constitutes this country's biggest drug problem. Children have to be told about alcohol in more not less detail and in different ways throughout their development, up to and including young adulthood.

What often happens is that schools try to prevent binge drinking by inviting a thirty-something 'former alcoholic' to tell pupils how being brutally addicted to alcohol destroyed their lives. Many students, by virtue of being young, feel invincible, and that being 'down and out' through alcohol isn't something that could happen to them. Children and even young adults simply don't think this far ahead and find it hard to identify with hypothetical eventualities fifteen or twenty years down the road. Therefore, they need to be made aware of the more everyday effects of alcohol on their bodies, lives and minds.

## Starting Age

Few parents, teachers or doctors I meet are aware of the position on when children should be able to taste their first drink of alcohol. The UK's Chief Medical Officer has made it explicitly clear that children should not drink any alcohol until they are at least fifteen years old; this includes parents offering alcohol to their children at home (which sits uncomfortably with the fact that parents can legally give children drinks from the age of five in their own home). In America, the Surgeon General has made it explicitly clear that no one should drink alcohol until they are at least twenty-one years old. This includes parents offering their twenty-year-olds a glass of wine at home. Both positions may well be unpopular among teenagers and young voters, and governments may lose sin tax as a result. So we have to ask ourselves why would the Chief

Medical Officer and Surgeon General make such unpopular decisions? Obviously, they are more than convinced that drinking alcohol earlier than the ages stipulated is far more likely to lead to alcohol problems, 'alcoholism' and harm to health.

The Chief Medical Officer (now retired) is deserving of a glowing legacy given Britain's adoration of alcohol, as behind the scenes his must have been a politically unpopular position to adopt. And while it probably would have been impossible for him to have recommended an even higher cut-off age, we must all reflect on the glaring six-year difference between Britain's and America's legal age for drinking. Doctors in both countries have obviously seen much of the same medical evidence, and there are certainly no medical studies indicating that British teenagers have different brains and livers from their American counterparts, so the reasons cannot be medical. However, Britain is so deeply entrenched in a drinking culture that it is a political necessity to allow discretionary drinking at a much younger age. To suggest that a sixteen- or seventeen-year-old must not drink would, it seems, be culturally and legally untenable (not to mention the reaction of the alcohol lobby). Yet there is no question that reducing exposure to alcohol until a much later age is effective in decreasing alcohol addiction, harm and deaths.

American states can make individual choices for many things, from the age of consent to the legal driving age, to marriage, as well as the death penalty. Yet despite each state's freedom of choice, while execution is one thing, alcohol is clearly another:

the US government has insisted that *all fifty states* raise their minimum legal drinking age to twenty-one. In addition, many states do not allow those under twenty-one even to enter an off-licence or a bar, and some do not allow anyone under twenty-one to drink anywhere, including home – in other words, alcohol consumption is banned completely before the age of twenty-one. And there has been no backlash as claimed by the alcohol lobby. In the 1970s the legal purchasing age was changed in the USA to eighteen and there was a notable increase in a range of alcohol-related problems, particularly road deaths. So individual states were later told by the federal government to raise the age back to twenty-one, and as they did so, state by state there was an obvious trend of reducing late-night traffic fatalities, as well as other problems. This has led to thousands of lives being saved each year from alcohol overdose and road-traffic accidents alone.

There is no magic bullet, and raising drinking ages does not cure an age-old problem. However, it certainly makes it less severe. Whether Britain and other countries want to raise the legal age at which children can be given alcohol at home, or buy (or be served) alcohol outside the home, should be a medical/child wellbeing issue, but it remains predominantly a political and financial one. At the moment, there is a glaring discrepancy between what is medically true and what we are told is officially true about what is in our children's best interests. And in order to protect our children in today's drinking culture, we really have to reconcile this disparity in competing truths.

## The Legal Loophole

In writing this paragraph, it has taken me three hours of telephone calls and emails to various alcohol-licensing authorities in local government to establish precisely what age and under what circumstances a teenager can swallow a pint in a pub. I finally spoke to the Home Office and this is what I found out:

- Under-fourteens can go into a pub with a children's certificate as long as they are with an adult – this could be a boyfriend or a 'mate' – and stay in the family room or garden, but they are not allowed to drink alcohol.
- Sixteen- to seventeen-year-olds can go into a pub and drink beer, wine and cider with a meal, as long as they are accompanied by another teenager who is eighteen years old and who buys the drinks.[7]

So while Britain bemoans her binge-drinking children, the laws regarding children and alcohol are a mess. It's time someone put two and two together: once a fourteen- or fifteen-year-old can get into a pub, it's going to be difficult to prevent them from swallowing someone else's alcohol, if not their own. The approach of allowing five-year-olds to drink alcohol at home, and underage teenagers to sit in pubs where older teenagers can easily buy them drinks, leaves just too much room for confusion and manoeuvre.

We should have a single legal drinking age, even if it is unenforceable. This will send a new message to parents and society about what is good for our children, and will make it easier to exert authority over those of them who increasingly feel entitled to drink. Enforcement agencies in the UK have been singularly unsuccessful in preventing off-licences, shops and pubs from selling alcohol to children. And children often feel (wrongly) entitled to be irritated when asked for verification of their age. The signs in shops politely saying, 'Please do not be offended if we ask for proof of age when you buy alcohol' should read: 'You MUST prove you are over eighteen or we will not sell you alcohol. Furthermore, if you try to buy it anyway, we'll call the police and beg them to arrest, jail and prosecute you.'

The House of Commons Health Committee (2010) positively savaged the impotence of government, saying: 'The Department for Culture, Media and Sport has shown extraordinary naivety in believing the Licensing Act 2003 would bring about "civilised café culture". In addition, the Act has failed to enable the local population to exercise adequate control of a licensing and enforcement regime which has been too feeble to deal with the problems it has faced.'

And when it comes to drinking in public, the laws seem to be at their most feeble. On the train last night, I was about to doze off when a news story caught my eye that just about sums it up. London's Camden Council and Camden Police have just announced their latest weapon in the fight against binge drinking: free flip-flops, lollipops, tea and biscuits, to be handed out to drunken revellers on Friday and Saturday nights. A spokesman for Camden Council explained that the scheme, called 'Camden Departure Lounge', is offering the free flip-flops for 'women struggling with high heels', while the drinks (tea, coffee, squash and water), biscuits and lollipops 'can stop people shouting, make people less aggressive and prevent post-alcohol hunger'. Camden's cabinet minister for community safety said they wanted to ensure that 'people who come to our area can enjoy themselves and get home safely'.[8]

Any measures to deal with drinking in public should have teeth not a gratuitous whimper. If a society allows young people to walk around town, beer in hand, there will be more trouble – pure and simple. It will also continue to normalise a sense of entitlement among people to drink as *they* see fit, not as the common good sees fit. Rather than offering free flip-flops to the inebriated, the co-enabling council should be giving them a free boot up the backside.

## The True Cost of the Price

Raising the age of both discretionary and legal drinking reduces addiction rates and harm for children. And so does raising the price of alcohol.

Many physicians and researchers around the world, including Britain's own Chief Medical Officer, now strongly believe that the government must raise the price of alcohol as a disincentive for both children and adults to buy it. A study in 2009 of 'Teenage drinking, alcohol availability and pricing' involving 9833 children in England concluded:

> Strategies to reduce alcohol-related harms in children should ensure bingeing is avoided entirely, address the excessively low cost of many alcohol products, and tackle the ease with which it can be accessed, especially outside of supervised environments ... While these measures are unlikely to eradicate the negative effects of alcohol on children, they may reduce them substantially while allowing children to prepare themselves for life in an adult environment dominated by this drug.[9]

Again, the House of Commons Health Committee makes brusque, unequivocal judgments, declaring that a rise in the price of alcohol was the most effective way of reducing the amount that people drink. They strongly recommend that the government introduces minimum pricing. And they dismiss as nonsense the false claims of the industry that raising the price won't change the degree of harmful drinking a society suffers as 'economic illiteracy. Different countries, like different people and groups, respond differently to price, but they all respond.'

They go on to obliterate the 'innocent victim' of price rises in alcohol. 'There is a myth widely propagated by parts of the drinks industry and politicians that a rise in prices would unfairly affect the majority of moderate drinkers.' They've done the maths and found that in sheer pounds and pence, precisely because someone *is* a moderate drinker, a price rise would have little effect on them. An increase to a minimum of 40 pence per unit, for example, 'would cost a moderate drinker 11p per week' extra.

Opponents of price increases claim that heavier drinkers are insensitive to price changes, but as a group their consumption will be most affected by price rises since they drink so much of the alcohol purchased in the country. (Ten per cent of the population drink 44 per cent of the alcohol consumed; 75 per cent of alcohol is drunk by people who exceed the recommended limits.) Minimum pricing would impact the most on those who drink cheap alcohol, in particular young binge drinkers and heavy low-income drinkers who suffer most from liver disease. It is estimated that a minimum price of 50p per unit would save over 3000 lives per year, and a minimum price of 40p, 1100 lives. 'Increasing the price of alcohol is thus the most powerful tool at the disposal of a government.'

For many, the entire concept of controlling access to and the price of alcohol seems draconian if not downright priggish. If people did drink 'responsibly' as the advert urges, and we didn't have such a problem with young drinking, government intervention of this type would indeed be nanny-state behaviour. However, we have a big alcohol problem – among our young, in particular – and so the state must be involved, just as it is in so many other areas involving substances which affect both collective good and individual welfare. The House of Commons Health Committee encapsulates our predicament perfectly in saying, 'Alcohol is no ordinary commodity, and its regulation is an ancient function of Government.' And let's not forget that the study in the *Lancet* in 2010 ranking twenty of Britain's most harmful drugs places alcohol at number one (see Chapter 1). We have to get used to the truth that alcohol is far more harmful than *all* popular illegal substances, including ecstasy, LSD and cannabis.[10]

## Marketing

The House of Commons Health Committee is entirely correct in saying that 'people have a right to know the risks they are running. Unfortunately, these campaigns are poorly funded.' The Committee compares this education poverty with expenditure on marketing by the drinks industry which was estimated to be between £600 and £800m in 2003 – 'The current system of controls on alcohol advertising and promotion is failing the young people it is intended to protect.'

Advertising and marketing has become extremely sophisticated in recent years with few controls in many countries. And it works, subversively, having a huge impact on the drinking habits of young people and enticing people to drink earlier and to a greater extent. Self-regulation guidelines have usually been ineffective and are violated by those they apply to.

And so the Committee has decided that as a matter of urgency alcohol advertising and promotion must be restricted in places where children are likely to be affected by it. 'The regulation of alcohol promotion should be completely independent of the alcohol and advertising industries.'

## Drink to the Future

All adults need to be aware that their favourite drug and social lubricant may have profound consequences when children indulge in it. The evidence is abundantly clear that ideally parents should not allow their children to consume alcohol at all – including having a drink at home – until they have reached at *least* the age of fifteen. As a health professional, however, I have to make it clear that on a purely health and development basis, *ideally* young people should not consume any alcohol at all until they have reached at least the age of twenty-four and a half. Therefore, the suggested minimum ages for drinking alcohol (fifteen) and our legal age for buying it (eighteen) have to be seen for what they are – cultural and political compromises, *not* accurate health guidelines. And parents, teachers and policy-makers alike must be especially aware of where information about alcohol education comes from.

Having said this, I am, despite the worrying information in this book, actually very optimistic about the future. Only a few decades ago it was perfectly acceptable to drink heavily and to drive your children around without seatbelts with a fag in your mouth. All that has now changed. In fact, Britain's drink–drive campaign is held up as one of the world's best ever examples of a successful health-education campaign. Similarly, sustained efforts to change attitudes towards public and passive smoking, and drinking while pregnant have also been far more successful than people ever expected. While the problems created by drinking, particularly in the young, may seem daunting at the moment, if we play our cards right, we have very good reasons to be cheerful.

And so let's raise our glasses to child sobriety and drink a toast to the exorcism of the Devil's buttermilk from our children's development.

# References

## Chapter 1


1 Health Education Authority, *Your Pocket Guide to Sex*, 1994.
2 Helen Brown, 'The celebrity guide to detox: Pass out, check in, and dry out', *Independent*, 6 January 2007.
3 Christina Valhouli, 'Your Health: Most Luxurious Places To Dry Out', *Forbes*, 2005.
4 Leah McLaren, 'Don't worry – drink and be merry', *Spectator*, 30 December 2009.
5 Justin Webb, 'Is Everyone in Britain Drunk?', *Daily Mail* , 8 July 2010.
6 John Humphrys, 'Drunken Anarchy', *Daily Mail*, 15 April 2010.
7 OECD, 'Doing Better for Children', 1 September 2009.
8 NHS Information Centre, 'Smoking, drinking and drug use among young people in England in 2008', 2009.
9 National Centre for Social Research, 'Smoking, drinking and drug use among young people in England: Findings by region, 2006 to 2008', January 2010.
10 Alcohol Concern, 'Right time, right place: alcohol-harm reduction strategies with children and young people', 2010.
11 Department of Health, 'Guidance on the Consumption of Alcohol by Children and Young People', 2009; Sir Liam Donaldson, Chief Medical Officer, 17 December 2009; Department of Health statistics, 22 March 2009; figures obtained by the Liberal Democrats and published in the British press: 'Teenage girls and alcohol poisoning', *Daily Telegraph*, 22 March 2009.
12 National Treatment Agency, 2007.

13 P. Quigley, et al., 'Toxicology: Prospective study of 101 patients with suspected drink spiking', *Emergency Medicine Australasia*, volume 21, pp. 222–8, 2009; Hywel Hughes, et al., 'A study of Patients Presenting to an Emergency Department Having Had a "Spiked Drink"', *Emergency Medicine Journal*, volume 24, pp. 89–91, 2007.
14 G. Li, et al., 'Use of alcohol as a risk factor for bicycling injury', *JAMA*, volume 285, pp. 893–6, 2001.
15 Coroners and Procurators Fiscal, 'Road Casualties Great Britain 2007. Department for Transport', September 2008.
16 Drinkaware, W. Johnson, as reported in the *Independent*: 'Third of under-24s "drink to get drunk"', 7 September 2010.
17 NHS Information Centre, Alcohol-related Admissions, October 2010.
18 House of Commons Health Committee, 'Alcohol', HC 151–1. London: The Stationery Office Limited, 2010.
19 ChildWise, 'Living with Alcohol', in conjunction with BBC, 5 July 2010.
20 British Liver Trust, 'Teen drink and drug use declining', as reported on BBC News, 23 July 2009.
21 Department for Children Schools and Families, 'Children, Young People and Alcohol', January 2010; also in the *Daily Mail* and *Evening Standard* (16 February 2010; 'Young people and alcohol', Department of Education, conducted by National Centre for Social Research, June 2010.)
22 'Alcohol In Europe. A Public Health Perspective: A report for the European Commission', Peter Anderson and Ben Baumberg, Institute of Alcohol Studies, UK, June 2006.
23 Directorate-General for Health and Consumer Protection, European Commission, 'Alcohol-Related Harm in Europe', Fact Sheet, 2006.
24 Royal College of Psychiatrists' Public Education Editorial Board, 'Alcohol and Depression', June 2008.
25 D. J. Nutt, et al., 'Drug harms in the UK: a multicriteria decision analysis', the *Lancet*, Early Online Publication, 1 November 2010, DOI:10.1016/S0140–6736(10)61462–6.
26 Royal College of Psychiatrists' Public Education Editorial Board, 'Alcohol: Our Favourite Drug', 2008.
27 Adam Edwards, 'Drinks are in the House: Has New Labour watered down Westminster's convivial drinking habits, asks Adam Edwards', telegraph.co.uk, 7 November 2002.
28 'MPs Drunk as They Voted on the Budget', *Daily Mail*, 12 July 2010; Tory MP 'too drunk to vote in Commons debate', *Daily Telegraph*, 12 July 2010; 'Mark Reckless MP sorry for being "too drunk to vote", BBC News, 11 July 2010.
29 Fact Sheet G19, General Series, the House of Commons Refreshment Department, House of Commons Information Office, 2003.

30 Alan Johnson, 'Nanny state, Nudge State or No state?', Speech by Rt Hon. Alan Johnson, MP, Secretary of State for Health RSA, Thursday 19 March 2009.
31 ancestry.co.uk (March 2010), http://www.ancestry.co.uk/about/default.aspx?section=pr-2010-03-27)
32 Bernard C. Peters, 'Hypocrisy on the Great Lakes Frontier: the Use of Whiskey by the Michigan Department of Indian Affairs', *Michigan Historical Review*, volume 18, issue 2, pp. 1–13, Fall 1992.
33 Thomas L. McKenney, *Sketches of a Tour to the Lakes of the Character of the Chippeway Indians, and the Incidents Connected with the Treaty of Fond Du Lac*, p. 197, Fielding Lucas, Jr, Baltimore, 1927.
34 O. Ned Eddins, 'Alcohol and the Indian Fur Trade', thefurtrapper.com, Afton, Wyoming, 2002.
35 Gore quoted in N. Thomas, *Discoveries: The Voyages of Captain Cook*, Penguin, 2003.
36 House of Commons Health Committee, 'Alcohol', HC 151–I. London: The Stationery Office Limited, 2010.
37 ibid.
38 ibid.
39 ibid.
40 ibid.
41 ibid.
42 ibid.
43 ibid.
44 Memorandum by the Royal College of Psychiatrists (AL49), 23 April 2009, Parliamentary Publications and Records.

## Chapter 2

1 Mark A. Bellis, et al., 'Teenage drinking, alcohol availability and pricing: a cross-sectional study of risk and protective factors for alcohol-related harms in school children', *BMC Public Health*, volume 9, p. 380, 2009, DOI:10.1186/1471-2458-9-380; Mark A. Bellis, et al., 'Predictors of risky alcohol consumption in schoolchildren and their implications for preventing alcohol-related harm', *Substance Abuse Treatment, Prevention, and Policy*, volume 2, p. 15, 2007, DOI:10.1186/1747-597X-2-15.
2 British Crime Survey, reported by Tom Whitehead in the *Daily Telegraph*, 'Number of "ladette women" fined for drunk and disorderly behaviour "rises by a third"', 14 June 2009; Tom Whitehead, 'Women and girls to blame for one in four violent attacks', *Daily Telegraph*, 26 May 2009.
3 Christine Griffin, et al., reported by Laura Donnelly in the *Daily*

*Telegraph*: 'Drunken celebrities are poor role models', 27 December 2008; study: Christine Griffin, et al., 'Branded Consumption and Social Identification, Young People and Alcohol' (RES-148-25-0021), Economic and Social Research Council (ESRC) project, 28 December 2008.

4 Christine Griffin, et al., 'The allure of belonging: Young people's drinking practices and collective identification', chapter in M. Wetherell (ed.), *Identity in the 21st Century: New Trends in New Times*, Palgrave, Macmillan, London, 2009.

5 A. Sigman, 'Visual Voodoo: The biological impact of watching television', *Biologist*, volume 54, issue 1, pp. 12–17, 2007; A. Sigman, 'A Source of Thinspiration: The biological landscape of media, body image and dieting', *Biologist*, volume 57, issue 3, pp. 117–21, 2010.

6 P. Anderson, et al., 'Impact of Alcohol Advertising and Media Exposure on Adolescent Alcohol Use: A Systematic Review of Longitudinal Studies', *Alcohol and Alcoholism*, volume 44, issue 3, pp. 229–43, 2009.

7 Reiner Hanewinkel and James D. Sargent, 'Longitudinal Study of Exposure to Entertainment Media and Alcohol Use Among German Adolescents', *Pediatrics*, volume 123, issue 3, pp. 989–995, 2009.

8 Reiner Hanewinkel, et al., 'Longitudinal study of parental movie restriction on teen smoking and drinking in Germany', *Addiction*, volume 103, issue 10, pp. 1722–30, 2008.

9 S. E. Tanski, et al., 'Parental R-rated Movie Restriction and Early-Onset Alcohol Use', *Journal of Studies on Alcohol and Drugs*, volume 71, issue 3, pp. 452–59, May 2010.

10 M. Stoolmiller, et al., 'R-rated Movie Viewing, Growth in Sensation Seeking and Alcohol Initiation: Reciprocal and Moderation Effects', *Prevention Science*, volume 11, pp. 1–13, 2010.

11 US Department of Health and Human Services, Surgeon General's Call to Action to Prevent and Reduce Underage Drinking, Office of the Surgeon General, 2007; B. F. Grant and D. A. Dawson, 'Age at onset of alcohol use and its association with DSM-IV alcohol abuse and dependence: results from the National Longitudinal Alcohol Epidemiologic Survey, *Journal of Substance Abuse*, volume 9, pp.103–10, 1997.

12 R. L. Collins, et al., 'Watching Sex on Television Predicts Adolescent Initiation of Sexual Behavior', *Pediatrics*, volume 114, issue 3, pp. e280–89, September 2004.

13 A. Chandra A, et al., 'Does Watching Sex on Television Predict Teen Pregnancy? Findings from a National Longitudinal Survey of Youth', *Pediatrics*, volume 122, issue 5, pp. 1047–54, November 2008.

14 Bruno Challier, et al., 'Associations of family environment and individual factors with tobacco, alcohol and illicit drug use in adolescents', *European Journal of Epidemiology*, volume 16, issue 1, pp. 33–42, 2000,

DOI 10.1023/A:1007644331197; R. A. Johnson, , et al., 'The relationship between family structure and adolescent substance use', Rockville, MD: Substance Abuse and Mental Health Services Administration, Office of Applied Studies, 1996.

15 M. Stern, J. E. Northman and M. R. Van Slyck, 'Father absence and adolescent "problem behaviors": alcohol consumption, drug use and sexual activity', *Adolescence*, Summer 1984, volume 19, issue 74, pp. 302–12; Dianna T. Kennya, Istvan Schreinera, 'Predictors of High-risk Alcohol Consumption in Young Offenders on Community Orders: Policy and Treatment Implications', *Psychology, Public Policy, and Law*, volume 15, issue 1, pp. 54–79, February 2009.

16 H. Sweeting, 'Teenage Family life, lifestyles and life chances', *International Journal of Law, Policy and the Family*, volume 12, issue 1, pp. 15–46, 1998.

17 M. Ely, et al., 'Teenage Family Life, Life Chances, Lifestyles and Health', *International Journal of Law, Policy and the Family*, volume 14, issue 1, pp. 1–30, April 2000.

18 Department for Children Schools and Families, 'Children, Young People and Alcohol', January 2010 (DCSF-RR195).

19 G. Ringbäck, et al., 'Mortality, severe morbidity, and injury in children living with single parents in Sweden: a population-based study', *Lancet*, volume 361, issue 9354, pp. 289–95, 25 January 2003.

20 Challier, op. cit.

21 G. Gallup Jr, 'What Americans think about their lives and families', *Families Magazine*, June 1982.

22 S. W. Sadava and M. M. Thompson, 'Loneliness, social drinking, and vulnerability to alcohol problems', *Canadian Journal of Behavioral Science*, volume 18, pp. 133–9, 1986.

23 Ingemar Åkerlind and Jan Olof Hörnquist, 'Loneliness and alcohol abuse: A review of evidences of an interplay', *Social Science & Medicine*, volume 34, issue 4, pp. 405–14, February 1992.

24 Michel F. Bonina, et al., 'Problem Drinking Behavior in Two Community-Based Samples of Adults: Influence of Gender, Coping, Loneliness', *Psychology of Addictive Behaviors*, volume 14, issue 2, pp. 151–61, June 2000.

25 NSPCC, 'ChildLine Casenotes: Children talking to ChildLine about loneliness', London, NSPCC, 2010; Cloke's comments appear on Mental Health Foundation http://www.mentalhealth.org.uk/media/news-releases/news-releases-2010/25-may-2010/).

26 Mental Health Foundation, 'The Lonely Society?', ISBN 978-1-906162-49-8, 2010.

27 A. Sigman, 'Well Connected? The biological implications of "social networking"', *Biologist*, volume 56, issue 1, pp. 14–20, 2009.

28 K. Abbassi, 'From The Editor: MMR and the value of word of mouth in social networks', *Journal of the Royal Society of Medicine*, volume 101, pp. 215–16, 2008.
29 ChildWise, Monitor Report 2007–2008.
30 Children's Society, 'Good Childhood Inquiry', published 25 February 2008, UK.
31 Sodexo-Times Higher Education University Lifestyle Survey, 'Survey finds it's all work, less play for the top-up generation', 11 September 2008.
32 Norman H. Nie, 'Sociability, Interpersonal Relations, and the Internet', *American Behavioral Scientist*, volume 45, issue 3, pp. 420–35, November 2001; Norman H. Nie, et al., 'Internet Use, Interpersonal Relations and Sociability: A Time Diary Study', in *The Internet in Everyday Life*, edited by Wellman and Haythornthwaite, Blackwell Publishers, 2003; Norman H. Nie, et al., 'Ten years after the birth of the Internet: how do Americans use the Internet in their daily lives?' report, Stanford University, 2005.
33 S. Konrath, et al., 'Empathy: College students don't have as much as they used to', University of Michigan News Service, 27 May 2010.
34 S. Konrath, et al. (under review), 'Changes in dispositional empathy in American college students over time: A meta-analysis', [also] Association for Psychological Science, 22nd Annual Convention, Boston, Friday 28 May, 2010.
35 ibid.
36 Comments by first author of study, Mary Helen Immordino-Yang, reported from news press release University of Southern California, 13 April 2009; Mary Helen Immordino-Yang, et al., 'Neural Correlates of Admiration and Compassion', *Proceedings of the National Academy of Sciences*, 2009; www.pnas.org.
37 R. Reich, 'The Future of Success', speech before the Los Angeles World Affairs Council, 21 February 2001.
38 Kathleen Christensen of the Alfred P. Sloan Foundation, as quoted in 'Families' Every Fuss, Archived and Analyzed', *New York Times*: Science, 22 May 2010.
39 Belinda Campos, et al., 'Opportunity for Interaction? A Naturalistic Observation Study of Dual-Earner Families After Work and School', *Journal of Family Psychology*, volume 23, issue 6, pp. 798–807, 2009.
40 J. Bradshaw, et al., 'An Index of Child Well-being in the European Union', *Social Indicators Research*, volume 80, issue 1, pp. 133–77, 2007.
41 S. S. Luthar, S. J. Latendresse, 'Children of the Affluent. Challenges to Well-Being', *Current Directions in Psychological Science*, volume 14, issue 1, pp. 49–53, 2005; the National Center on Addiction and Substance Abuse at Columbia University, 'The Importance of Family Dinners.

National Center on Addiction and Substance Abuse at Columbia', 2005, statements reported by Fox News – Health: 'Family Meals Help Teens Avoid Smoking, Alcohol, Drugs', www.foxnews.com/story.

## Chapter 3

1 Susan Bell, 'French the worst for smacking children', *Scotsman*, 8 December.
2 Directorate-General for Health and Consumer Protection, European Commission, Fact Sheet, 'Alcohol-Related Harm in Europe', 2006.
3 J. Hatton, et al., 'Drinking patterns, dependency and life-time drinking history in alcohol-related liver disease', *Addiction*, volume 104, pp. 587–92, 2009.
4 Department for Children, Schools and Families (RR195), 'Children, Young People and Alcohol', January 2010; also in the *Daily Mail* and *Evening Standard*.
5 Stephen J. Bahr, John P. Hoffmann, 'Parenting Style, Religiosity, Peers, and Adolescent Heavy Drinking', *Journal of Studies on Alcohol and Drugs*, volume 71, pp. 539–43, 2010.
6 S. M. Ryan, et al., 'Parenting factors associated with reduced adolescent alcohol use: a systematic review of longitudinal studies', *Australia and New Zealand Journal of Psychiatry*, volume 44, issue 9, pp. 774–83, September 2010.
7 J. D. Sellman, et al., 'How to reduce alcohol-related problems in adolescents: what can parents do and what can the government do?', volume 44, issue 9, pp. 771–3, September 2010.
8 a. Caitlin Abar, et al., 'Myth of the forbidden fruit: The impact of parental modeling and permissibility on college alcohol use and consequences', May 2009; b. In Caitlin Abar and R. Turrisi (Chairs), 'Putting the "European Drinking Model" to the Test in the US: Evidence and Implications for Prevention', symposium presented at the annual meeting of the Society for Prevention Research, Washington, DC; c. Caitlin Abar, et al., 'The Impact of Parental Modeling and Permissibility on Alcohol Use and Experienced Negative Drinking Consequences in College', *Addictive Behaviors*, volume 34, issue 6–7, pp. 542–7, 2009.
9 K. Foley, et al., 'Adults' approval and adolescents' alcohol use', *Adolescent Health*, volume 35, issue 4, pp. 345.e17–345.e26, October 2004.
10 T. A. Walls, et al., 'Parents do matter: a longitudinal two-part mixed model of early college alcohol participation and intensity', *Journal of Studies on Alcohol and Drugs*, volume 70, issue 6, pp. 908–18, November 2009.

11 R. Turrisi and A. E. Ray, 'Sustained parenting and college drinking in first-year students', *Developmental Psychobiology*, volume 52, issue 3, pp. 286–94, March 2010.

12 S. Mrug, et al., 'School-Level Substance Use: Effects on Early Adolescents' Alcohol, Tobacco, and Marijuana Use', *Journal of Studies on Alcohol and Drugs*, volume 71, pp. 488–95, 2010.

13 K. A. Komro, et al., 'Effects of home access and availability of alcohol on young adolescents' alcohol use', *Addiction*, volume 102, pp. 1597–1608, 2007; DOI: 10.1111/j.1360-0443.2007.01941.x.

14 H. van der Vorst, et al., 'Do parents and best friends influence the normative increase in adolescents' alcohol use at home and outside the home?', *Journal of Studies on Alcohol and Drugs*, volume 71, issue 1, pp. 105–14, 2010; and author interview: 'Teens Who Drink With Parents May Still Develop Alcohol Problems', *Science Daily*, 2 February 2010, http://www.sciencedaily.com/releases/2010/01/100127095930.htm)

15 Nikolaus Koutakis, et al., 'Reducing youth alcohol drinking through a parent-targeted intervention: the Örebro Prevention Program', *Addiction*, volume 103, pp. 1629–37, 2009.

16 S. Madon, et al., 'The mediation of mothers' self-fulfilling effects on their children's alcohol use: self-verification, informational conformity, and modeling processes', *Journal of Personality and Social Psychology*, volume 95, issue 2, pp. 369–84, August 2008.

17 Mennella, et al., 'Children's hedonic responses to the odors of alcoholic beverages: A window to emotions', *Alcohol*, volume 42, issue 4, pp. 249–60, 2008.

18 R. Alati, et al., 'In utero alcohol exposure and prediction of alcohol disorders in early adulthood: A birth cohort study', *Archives of General Psychiatry*, volume 63, pp. 1009–16, 2006; J. S. Baer, et al., 'A 21-year longitudinal analysis of the effects of prenatal alcohol exposure on young adult drinking', *Archives of General Psychiatry*, volume 60, pp. 377–86, 2003.

19 S. L. Youngentob and J. I. Glendinning, 'Fetal Exposure to Ethanol Increases Postnatal Acceptance by Altering its Odor and Taste', *Proceedings of the National Academy of Sciences*, volume 106, pp. 5359–64, 2009.

20 Amber M. Eade, et al., 'The consequence of fetal ethanol exposure and adolescent odor re-exposure on the response to ethanol odor in adolescent and adult rats', *Behavioral and Brain Functions*, 5:3, 2009.

21 Steven L. Youngentob, Ph.D., web page: 'In Utero Ethanol Exposure, Chemosensory Plasticity and Adolescent Drug Preference', http://www.upstate.edu/neurophys/faculty.php?EmpID=IUGIEJPx,2010.

22 Eilidh M. Duncan, et al., 'Assessing the potential impact of alcohol use during pregnancy on women in the postpartum period', presentation at British Psychological Society annual conference, Brighton, 2–4 April 2009.

23 I. Nulman, et al., 'Binge alcohol consumption by non-alcohol-dependent women during pregnancy affects child behaviour, but not general intellectual functioning; a prospective controlled study', *Archives of Women's Health*, volume 7, issue 3, pp. 173–81, 2004.
24 K. Sayal, et al., 'Binge Pattern of Alcohol Consumption During Pregnancy and Childhood Mental Health Outcomes: Longitudinal Population-Based Study', *Pediatrics*, volume 123, issue 2, pp. e289-e296, 2009.
25 Brian M. D'Onofrio, 'Causal Inferences Regarding Prenatal Alcohol Exposure and Childhood Externalizing Problems', *Archives of General Psychiatry*, volume 64, issue 11, pp. 1296–1304, 2007.
26 D. A. Dawson, et al., 'Age at First Drink and the First Incidence of Adult-Onset DSM-IV Alcohol Use Disorders', *Alcoholism: Clinical & Experimental Research*, volume 32, issue 12, pp. 2149–60, December 2008.
27 D. J. DeWit, et al., 'Age at first alcohol use: a risk factor for the development of alcohol disorders', *American Journal of Psychiatry*, volume 157, pp. 745–50, 2000.
28 A. Agrawal, et al. (2009), 'Evidence for an interaction between age at first drink and genetic influences on DSM-IV alcohol dependence symptoms', *Alcoholism: Clinical & Experimental Research*, volume 33, issue 12, pp. 2047–56, 2009; 'Young age at first drink may affect genes, alcoholism risk', News Release, Washington University School of Medicine, 30 September 2009.
29 Antoniette M. Maldonado-Devincci, et al., 'Alcohol during adolescence selectively alters immediate and long-term behavior and neurochemistry', *Alcohol*, volume 44, issue 1, pp. 57–66, January 2010.
30 Consuelo Guerri and María Pascual, 'Mechanisms involved in the neurotoxic, cognitive, and neurobehavioral effects of alcohol consumption during adolescence', *Alcohol*, volume 44, issue 1, pp. 15–26, January 2010.
31 US Department of Health and Human Services, National Institutes of Health, 'Make a Difference', NIH Publication number 06–4314, 2009.
32 Department of Health, 'Guidance on the consumption of alcohol by children and young people. A report by the Chief Medical Officer' (Sir Liam Donaldson), 17 December 2009.
33 US Department of Health and Human Services, 'The Surgeon General's Call to Action to Prevent and Reduce Underage Drinking', US Department of Health and Human Services, Office of the Surgeon General, 2007.
34 Alcohol Concern statement, 'Call to stop children's drinking', reported by BBC News, 27 April 2007; http://news.bbc.co.uk/1/hi/uk/6596515.stm.
35 Ali Bloom, 'Under age, over the limit, on school premises – now in court', published in the *Times Education Supplement: Analysis*, 26 June 2009.

36 M. A. Enoch, 'Genetic and Environmental Influences on the Development of Alcoholism: Resilience vs. Risk', *Annals of New York Academy of Sciences*, volume 1094, pp. 193–201, 2006.
37 ibid.
38 S. C. Wilsnack, et al., 'Childhood sexual abuse and women's substance abuse: national survey findings', *Journal of Studies on Alcohol and Drugs*, volume 58, pp. 264–71, 1997.
39 R. S. Trim, et al., 'The Relationships of the Level of Response to Alcohol and Additional Characteristics to Alcohol Use Disorders across Adulthood: A Discrete-Time Survival Analysis', *Alcoholism: Clinical & Experimental Research*, volume 33, issue 9, pp. 1562–70, September 2009; and Trim, et al., press release by the journal entitled, 'A person's high or low response to alcohol says much about their risk for alcoholism', 22 May 2009.
40 A. Webb, et al., 'The Investigation into CYP2E1 in Relation to the Level of Response to Alcohol Through a Combination of Linkage and Association Analysis', *Alcoholism: Clinical & Experimental Research*, Early Online: 19 October 2010; DOI: 10.1111/j.1530-0277.2010.01317.x.
41 G. Joslyn, et al., 'Human Variation in Alcohol Response Is Influenced by Variation in Neuronal Signaling Genes', *Alcoholism: Clinical & Experimental Research*, volume 34, issue 5, pp. 800–12, 2010.
42 B. Tabakoff, et al., 'Study on State and Trait Markers of Alcoholism. Genetical genomic determinants of alcohol consumption in rats and humans', *BMC Biology*, volume 7, p. 70, 2009; http://www.sciencedaily.com/releases/2009/10/091026192900.htm

## Chapter 4

1 US Department of Health and Human Services, 'The Surgeon General's Call to Action to Prevent and Reduce Underage Drinking', US Department of Health and Human Services, Office of the Surgeon General, 2007.
2 Consuelo Guerri and María Pascual, 'Mechanisms involved in the neurotoxic, cognitive, and neurobehavioral effects of alcohol consumption during adolescence', *Alcohol* (Fayetteville, NY), volume 44, issue 1, pp. 15–26, 1 January 2010; DOI: 10.1016/j.alcohol.2009.10.003.
3 US Department of Health and Human Services, 'Tenth Special Report to Congress on Alcohol and Health', 2000, chapter 2: Alcohol and the Brain: The Neurotoxicity of Alcohol, pp. 134–42.
4 A. Biller, et al., 'The effect of ethanol on human brain metabolites longitudinally characterized by proton MR spectroscopy', *Journal of Cerebral Blood Flow & Metabolism*, volume 29, pp. 891–902, 2009.

5 R. J. Ward, et al., 'Biochemical and Neurotransmitter Changes Implicated in Alcohol-Induced Brain Damage in Chronic or "Binge Drinking" Alcohol Abuse', *Alcohol & Alcoholism*, volume 44, issue 2, pp. 128–35, 2009.

6 T. McQueeny, et al., 'Altered White Matter Integrity in Adolescent Binge Drinkers', *Alcoholism: Clinical & Experimental Research*, volume 33, issue 7, pp. 1278–85, 2009.

7 S. Tapert S, as interviewed and reported on National Public Radio (NPR), 'Teen Drinking May Cause Irreversible Brain Damage', by Michelle Trudeau, 25 January 2010.

8 ibid.

9 L. M. Squeglia, et al., 'Initiating moderate to heavy alcohol use predicts changes in neuropsychological functioning for adolescent girls and boys', *Psychology of Addictive Behaviors*, volume 23, issue 4, pp. 715–22, December 2009.

10 S. Tapert, transcript from National Public Radio (NPR), 'With Drinking, Parent Rules Do Affect Teens' Choices', heard on Morning Edition, 31 May 2010.

11 B. J. Nagel, et al., 'Reduced hippocampal volume among adolescents with alcohol use disorders without psychiatric comorbidity', *Psychiatry Research*, volume 139, issue 3, pp. 181–90, 30 August 2005.

12 K. L. Medina, 'Effects of alcohol and combined marijuana and alcohol use during adolescence on hippocampal volume and asymmetry', *Neurotoxicology and Teratology*, volume 29, issue 1, pp. 141–52, January–February 2007.

13 M. D. De Bellis, et al., 'Hippocampal Volume in Adolescent-Onset Alcohol Use Disorders', *American Journal of Psychiatry*, volume 157, pp. 737–44, 2000.

14 A. D. Schweinsburg, et al., 'A preliminary study of functional magnetic resonance imaging response during verbal encoding among adolescent binge drinkers', *Alcohol*, volume 44, issue 1, pp. 111–17, February 2010.

15 M. A. Taffe, et al., 'Long-lasting reduction in hippocampal neurogenesis by alcohol consumption in adolescent non-human primates', *Proceedings of the National Academy of Sciences*, volume 107, pp. 11104–9, 2010.

16 R. J. Ward, et al., 'Neuro-inflammation induced in the hippocampus of "binge-drinking" rats may be mediated by elevated extracellular glutamate content', *Journal of Neurochemistry*, volume 111, issue 5, pp. 1119–28, December 2009.

17 C. A. Paul, et al., 'Association of Alcohol Consumption With Brain Volume in the Framingham Study', *Archives of Neurology*, volume 65, issue 10, p. 1363, 2008.

18 K. L. Medina, et al., 'Prefrontal Cortex Volumes in Adolescents with

Alcohol Use Disorders: Unique Gender Effects', *Alcoholism: Clinical & Experimental Research*, volume 32, issue 3, pp. 386–94, March 2008.

19 M. D. De Bellis, et al., 'Prefrontal Cortex, Thalamus, and Cerebellar Volumes in Adolescents and Young Adults with Adolescent-Onset Alcohol Use Disorders and Comorbid Mental Disorders', *Alcoholism: Clinical & Experimental Research*, volume 29, issue 9, pp. 1590–1600, 2005.

20 Bernd Figner, et al., 'Lateral prefrontal cortex and self-control in intertemporal choice', *Nature Neuroscience*, volume 13, pp. 538–9, 2010.

21 Shamay-Tsoory, et al., 'The role of the orbitofrontal cortex in affective theory of mind deficits in criminal offenders with psychopathic tendencies', *Cortex*, volume 46, issue 5, p. 668, 2010.

22 N. A. Shamosh, et al., 'Individual differences in delay discounting: relation to intelligence, working memory, and anterior prefrontal cortex', *Psychological Science*, volume 19, issue 9, pp. 904–11, September 2008.

23 H. R. White, et al., 'Associations Between Heavy Drinking and Changes in Impulsive Behavior Among Adolescent Boys', *Alcoholism: Clinical & Experimental Research*, DOI: 10.1111/j.1530-0277.2010.01345.x; and associated press release: 'Impulsive behavior in males increases after periods of heavy drinking', *Alcoholism: Clinical & Experimental Research*, 16 November 2010.

24 M. D. De Bellis, et al., 'Diffusion Tensor Measures of the Corpus Callosum in Adolescents with Adolescent Onset Alcohol Use Disorders', *Alcoholism: Clinical & Experimental Research,* volume 32, issue 3, pp. 395–404, March 2008.

25 W. J. McBride, et al., 'Changes in gene expression in regions of the extended amygdala of alcohol-preferring rats after binge-like alcohol drinking', *Alcohol*, volume 44, issue 2, pp. 171–83, March 2010.

26 Cindy L. Ehlers and José R. Criado, 'Adolescent ethanol exposure: does it produce long-lasting electrophysiological effects?', *Alcohol*, volume 44, issue 1, pp. 27–37, January 2010.

27 E. Tasali E, et al., 'Slow-wave sleep and the risk of type 2 diabetes in humans', *Proceedings of the National Academy of Sciences*, volume 105, issue 3, pp. 1044–49, 22 January 2008.

28 Linda Patia Spear and Elena I. Varlinskaya, 'Sensitivity to ethanol and other hedonic stimuli in an animal model of adolescence: Implications for prevention science?', *Developmental Psychobiology*, volume 52, pp. 236–43, 2010.

29 US Department of Health, 'Alcohol And Other Drug Use And Academic Achievement', 2009.

30 National Centre for Social Research, 'Young people & alcohol', by Department for Children, Schools and Families, project number P2958, June 2010.

31 M. López-Frías, 'Alcohol consumption and academic performance in a population of Spanish high school students', *Journal of Studies on Alcohol and Drugs*, volume 62, issue 6, pp. 741–4, November 2001.

32 R. W. Hingson, 'Impact of Alcohol Use on School Performance of US High School Students', American Public Health Association Scientific Session,135th Annual meeting, Washington DC, 5 November 2007.

33 J. G. Bachman, et al., 'The Education–Drug Use Connection: How Successes and Failures in School Relate to Adolescent Smoking, Drinking, Drug Use, and Delinquency', Lawrence Erlbaum Associates/Taylor & Francis, 2007.

34 University of Minnesota Boynton Health Service, College Student Health Survey, 20 October 2008.

35 D. L. Thombs, et al., 'Undergraduate Drinking and Academic Performance: A Prospective Investigation With Objective Measures', *Journal of Studies on Alcohol and Drugs*, volume 70, pp. 776–85, 2009.

## Chapter 5

1 A. Schützwohl, et al., 'How Willing Are You to Accept Sexual Requests from Slightly Unattractive to Exceptionally Attractive Imagined Requestors?', *Human Nature* volume 20, pp. 282–93, 2009.

2 R. D. Clark and E. Hatfield, 'Gender differences in receptivity to sexual offers', *Journal of Psychology and Human Sexuality*, volume 2, pp. 39–54, 1989.

3 L.G. Halsey, et al., 'An explanation for enhanced perceptions of attractiveness after alcohol consumption', *Alcohol,* volume 44, issue 4, pp. 307–13, June 2010.

4 Vincent Egan and Giray Cordan, 'Barely legal: Is attraction and estimated age of young female faces disrupted by alcohol use, make up, and the sex of the observer?', *British Journal of Psychology*, volume 100, issue 2, pp. 415–27, May 2009.

5 WHO, 'Alcohol Use and Sexual Risk Behaviour: A Cross-Cultural Study In Eight Countries', Geneva, 2005.

6 M. A. Bellis, et al., 'Sexual uses of alcohol and drugs and the associated health risks: A cross-sectional study of young people in nine European cities', *BMC Public Health*, volume 8, p. 155, 2008.

7 K. Standerwick, et al., 'Binge drinking, sexual behaviour and sexually transmitted infection in the UK', *International Journal of STD & AIDS*, volume 18, pp. 810–13, 2007.

8 Jennie Connor, 'Drinking history, current drinking and problematic sexual experiences among university students', *Australian and New Zealand Journal of Public Health*, volume 34, issue 5, pp. 487–94, October 2010.

9 Paul A. Agius, 'Sexual behaviour and related knowledge among a representative sample of secondary school students between 1997 and 2008', *Australian and New Zealand Journal of Public Health*, volume 34, issue 5, pp. 476–81, October 2010.
10 J. M. Boden, et al., 'Alcohol and STI risk: Evidence from a New Zealand longitudinal birth cohort', *Drug and Alcohol Dependence*, volume 113, issues 2–3, pp. 200–6, 2011.
11 C. R. Aicken, et al., 'Alcohol misuse, sexual risk behaviour and adverse sexual health outcomes: evidence from Britain's national probability sexual behaviour surveys', *Journal of Public Health*, first published online 12 August 2010.
12 Hywel Hughes, et al., 'A Study of Patients Presenting to an Emergency Department Having Had a "Spiked Drink"', *Emergency Medicine Journal*, volume 24, pp. 89–91, 2007.
13 P. Quigley, et al., 'TOXICOLOGY: Prospective study of 101 patients with suspected drink spiking', *Emergency Medicine Australasia*, volume 21, pp. 222–8, 2009.
14 A. Burgess, et al., 'Embodying Uncertainty? Understanding Heightened Risk Perception of Drink "Spiking"', *British Journal of Criminology*, volume 49, issue 6, pp. 848–62, 2009.
15 A. M. White, et al., 'Prevalence and correlates of alcohol-induced blackouts among college students: Results of an e-mail survey', *Journal of American College Health*, volume 51, pp. 117–31, 2002.
16 M. S. Mumenthaler, et al., 'Gender differences in moderate drinking effects', *Alcohol Research & Health*, volume 23, pp. 55–64, 1999.
17 Mark A. Bellis, et al., 'Teenage drinking, alcohol availability and pricing: a cross-sectional study of risk and protective factors for alcohol-related harms in school children', *BMC Public Health*, volume 9, p. 380, 2009; DOI:10.1186/1471-2458-9-380.
18 US Department of Health and Human Services, 'Alcohol's Damaging Effects on the Brain', *Alcohol Alert*, number 63, October 2004.
19 K. Mann, et al., 'Neuroimaging of Gender Differences in Alcohol Dependence: Are Women More Vulnerable?' *Alcoholism: Clinical & Experimental Research*, volume 29, issue 5, pp. 896–901, May 2005.
20 B. Flannery, et al., 'Gender Differences in Neurocognitive Functioning Among Alcohol-Dependent Russian Patients', *Alcoholism: Clinical & Experimental Research*, volume 31, issue 5, pp. 745–54, May 2007.
21 'Alcohol and Depression', Royal College of Psychiatrists Public Education Editorial Board, June 2008.
22 M. Frezza, et al., 'High blood alcohol levels in women. The role of decreased gastric alcoholdehydrogenase activity and first-pass metabolism', *New England Journal of Medicine*, volume 322, issue 2, pp. 95–9,

11 January 1990; G. Pozzato, et al., 'Ethanol metabolism and aging: the role of "first pass metabolism" and gastric alcohol dehydrogenase activity', *Journals of Gerontology. Series A, Biological Sciences and Medical Sciences*, volume 50, issue 3, pp. B135–41, May 1995.

23 Enrique Baraona, et al., 'Gender Differences in Pharmacokinetics of Alcohol', *Alcoholism: Clinical & Experimental Research*, volume 25, issue 4, pp. 502–7, April 2001.

24 C. P. Day, 'Who gets alcoholic liver disease: nature or nurture?' *Journal of the Royal College of Physicians*, London, volume 34, issue 6, pp. 557–62, November–December 2000.

25 Mary Ann Emanuele, MD, et al., 'Alcohol's Effects on Female Reproductive Function', National Institute on Alcohol Abuse and Alcoholism, June 2003.

26 M. H. Kaufman, 'The Teratogenic Effects of Alcohol Following Exposure During Pregnancy, and Its Influence on the Chromosome Constitution of the Pre-Ovulatory Egg', *Alcohol & Alcoholism*, volume 32, issue 2, pp. 113–28, 1997.

27 Lord Whaddon, 'Alcohol abuse', transcript of discussion held at the House of Lords, Hansard 290, number 41, p. 662, 1987.

28 N. Kaminen-Ahola, et al., 'Maternal Ethanol Consumption Alters the Epigenotype and the Phenotype of Offspring in a Mouse Model', *Public Library of Science: Genetics*, volume 6, issue 1, p. e1000811, 2010; DOI:10.1371/journal.pgen.1000811.

29 M. E. Pembrey, et al., 'Sex-specific, male-line transgenerational responses in humans', *European Journal of Human Genetics*, volume 14, pp. 159–66, 2006.

30 J. Willford, et al. (in press), 'Caudate asymmetry: A neurobiological marker of moderate prenatal alcohol exposure in young adults', *Neurotoxicology and Teratology*.

31 C. S. Berkey, et al., 'Prospective study of adolescent alcohol consumption and risk of benign breast disease in young women', *Pediatrics*, volume 125, issue 5, pp. e1081–7, 2010; DOI:10.1542/peds.2009–2347.

32 Joseph W. LaBrie, et al., 'What Men Want: The Role of Reflective Opposite-Sex Normative Preferences in Alcohol Use Among College Women', *Psychology of Addictive Behaviors*, volume 23, issue 1, 2009.

## Chapter 6

1 Adult Obesity and Overweight in the European Union (EU27), International Association for the Study of Obesity, 2010.

2 ibid.

3 OECD, 'Obesity and the Economics of Prevention: Fit not Fat', United Kingdom (England) Key Facts, 23 September 2010.
4 ibid.
5 Standard Life, 'Britain Heading Towards Alcohol Obesity', 23 August 2007.
6 Cynthia Ma and Aminah Jatoi, 'Wine for Appetite Loss: "How Do You Know?"', *Journal of Clinical Oncology*, volume 25, number 10, pp. 1285–7, 1 April 2007.
7 E. Jequier, 'Alcohol intake and body weight: A paradox', *American Journal of Clinical Nutrition*, volume 69, pp. 173–4, 1999.
8 M. S. Westerterp-Plantenga and C. R. Verwegen, 'The appetizing effect of an aperitif in overweight and normal-weight humans', *American Journal of Clinical Nutrition*, volume 69, pp. 205–12, 1999.
9 M. R. Yeomans, et al., 'Alcohol and food intake', *Current Opinion in Clinical Nutrition & Metabolic Care*, volume 6, pp. 639–44, 2003.
10 Standard Life, op. cit.
11 Rosalind A. Breslow and Barbara A. Smothers, 'Drinking Patterns and Body Mass Index in Never Smokers', *American Journal of Epidemiology*, volume 161, issue 4, pp. 368–76, 2005.
12 A. Raben, et al., 'Meals with similar energy densities but rich in protein, fat, carbohydrate, or alcohol have different effects on energy expenditure and substrate metabolism but not on appetite and energy intake', *American Journal of Clinical Nutrition*, volume 77, pp. 91–100, 2003.
13 L. Haines, et al., 'Rising Incidence of Type 2 Diabetes in Children in the UK', *Diabetes Care*, volume 30, pp. 1097–1101, May 2007.
14 U. Risérus and E. Ingelsson, 'Alcohol Intake, Insulin Resistance, and Abdominal Obesity in Elderly Men', *Obesity*, volume 15, pp. 1766–73, 2007.
15 BAAPS: British Association of Aesthetic Plastic Surgeons, 'Britons over the moob: male breast reduction nearly doubles in 2009', press release, London, UK, 1 February 2010.
16 BBC, 'Male breast ops "increase by 80 per cent"', 1 February 2010.
17 USC, 'Gynecomastia', doctors of USC, 2010.
18 'Alcohol and Depression', Royal College of Psychiatrists' Public Education Editorial Board, 2008.
19 S. Oesterle, et al., 'Adolescent Heavy Episodic Drinking Trajectories and Health in Young Adulthood', *Journal of Studies on Alcohol and Drugs*, volume 65, issue 2, pp. 204–12, March 2004.
20 D. S. Gaur, et al., 'Alcohol intake and cigarette smoking: Impact of two major lifestyle factors on male fertility', *Indian Journal of Pathology and Microbiology*, volume 53, pp. 35–40, 2010.
21 M. Hasselblatt, et al., 'Persistent disturbance of the hypothalamic–pituitary–gonadal axis in abstinent alcoholic men', *Alcohol and Alcoholism*, volume 38, issue 3, pp. 239–42, 2003.

22 Lee Ack, et al., 'The effect of alcohol drinking on erectile dysfunction in Chinese men', *International Journal of Impotence Research*, volume 22, pp. 272–8, July/August 2010.

23 NHS, 'Common questions about alcohol', UK Crown Copyright 2010.

24 Carol A. Derby, et al., 'Modifiable risk factors and erectile dysfunction: can lifestyle changes modify risk?', *Urology*, volume 56, issue 2, pp. 302–6, August 2000.

25 A. B. Araujo, et al., 'Relation between Psychosocial Risk Factors and Incident Erectile Dysfunction: Prospective Results from the Massachusetts Male Aging Study', *American Journal of Epidemiology*, volume 152, issue 6, pp. 533–41, 2000.

26 L. Cordain, et al., 'Acne Vulgaris: A Disease of Western Civilization', *Archives of Dermatology*, volume 138, pp. 1584–90, 2002.

27 E. M. Higgins and A. W. du Vivier, 'Cutaneous disease and alcohol misuse', *British Medical Bulletin*, volume 50, issue 1, pp. 85–98, January 1994.

28 'Consideration of the Anabolic Steroids', Advisory Council on the Misuse of Drugs, September 2010.

29 The American Council on Exercise, 'Fit Facts: Alcohol eats away at muscle mass', 2009.

30 American College of Sports Medicine, 'Position Statement on: The Use of Alcohol in Sports', 2010; Stella Lucia Volpe, 'Alcohol and Athletic Performance', *ACSM's Health and Fitness Journal*, volume 14, issue 3, pp. 28–30, 2010.

31 M. Hoxha, et al., 'Shortened telomeres in subjects with heavy alcohol consumption', paper presented to American Association for Cancer Research Annual Conference, 21 April 2010.

32 Leonardo Pignataro, et al., 'Alcohol Regulates Gene Expression in Neurons via Activation of Heat Shock Factor', *Journal of Neuroscience*, volume 27, pp. 12957–66, November 2007.

33 R. A. Kloner and S. H. Rexkall, 'To drink or not to drink? That is the question', *Circulation*, volume 116, pp. 1306–17, 2007; J. Rehm, et al., 'The relationship of average volume of alcohol consumption and patterns of drinking to burden of disease: an overview', *Addiction*, volume 98, pp. 1209–28, 2003.

34 Giuseppe Lippi, 'Red wine and cardiovascular health: the "French Paradox" revisited', *International Journal of Wine Research*, volume 2, pp. 1–7, 2010.

35 Johan Jarl, Ulf G. Gerdtham and Klara Hradilova Selin, 'Medical net cost of low alcohol consumption – a cause to reconsider improved health as the link between alcohol and wage?', *Cost Effectiveness and Resource Allocation*, volume 7, p. 17, 23 October 2009.

36 P. Fenoglio, et al., 'The Social Cost of Alcohol, Tobacco and Illicit Drugs in France, 1997', *European Addiction Research*, volume 9, issue 1, pp. 18–28, 2003.

37 Peter Anderson and Ben Baumberg, 'ALCOHOL IN EUROPE A PUBLIC HEALTH PERSPECTIVE: A report for the European Commission', Institute of Alcohol Studies, UK, June 2006.

38 F. D. Fuchs, et al., 'Association between Alcoholic Beverage Consumption and Incidence of Coronary Heart Disease in Whites and Blacks. The Atherosclerosis Risk in Communities Study', *American Journal of Epidemiology*, volume 60, pp. 466–74, 2004.

39 T. S. Naimi, et al., 'Cardiovascular risk factors and confounders among non-drinking and moderate-drinking US adults', *American Journal of Preventive Medicine*, volume 28, pp. 369–73, 2005.

40 B. Rodgers, et al., 'Non-linear relationships in associations of depression and anxiety with alcohol use', *Psychological Medicine*, volume 30, issue 2, pp. 421–32, 2000.

41 T. K. Greenfield, et al., 'Effects of depression and social integration on the relationship between alcohol consumption and all-cause mortality', *Addiction*, volume 97, pp. 29–38, 2002.

42 Professor Ian Gilmore, House of Commons, Minutes of Evidence Taken Before Health Committee, Thursday 23 April 2009, Uncorrected Transcript of Oral Evidence.


## Chapter 7


1 D. M. Fergusson, et al., 'Tests of Causal Links Between Alcohol Abuse or Dependence and Major Depression', *Archives of General Psychiatry*, volume 66, issue 3, pp. 260–6, 2009.

2 D. M. Fergusson, et al., 'Structural models of the comorbidity of internalizing disorders and substance use disorders in a longitudinal birth cohort', *Social Psychiatry and Psychiatric Epidemiology*, 0.1007/s00127-010-0268-1, 8 July 2010.

3 Z. A. Rodd, et al., 'The reinforcing actions of a serotonin-3 receptor agonist within the ventral tegmental area: evidence for subregional and genetic differences and involvement of dopamine neurons', *Journal of Pharmacology and Experimental Therapeutics*, volume 321, pp. 1003–12, 2007.

4 K. K. Szumlinski, et al., ' Accumbens neuro-chemical adaptations produced by binge-like alcohol consumption', *Psychopharmacology* (Berlin), volume 190, pp. 415–31, 2007.

5 J. L. Wang and S. B. Patten, 'Alcohol Consumption and Major Depression: Findings from a Follow-Up Study', *Canadian Journal of Psychiatry*, number 46, pp. 632–8, 2001.

6 K. Graham, et al., 'Does the Association Between Alcohol Consumption and

Depression Depend on How They Are Measured?' *Alcoholism: Clinical & Experimental Research*, volume 31, issue 1, pp. 78–88, January 2007.

7 'Alcohol and Depression', Royal College of Psychiatrists Public Education Editorial Board, June 2008.

8 D. S. Pine, et al., 'Adolescent Depressive Symptoms as Predictors of Adult Depression: Moodiness or Mood Disorder?', *American Journal of Psychiatry*, volume 156, pp. 133–5, January 1999; *American Journal of Psychiatry*, volume 159, issue 7, pp. 1235–7, July 2002; T. Aalto-Setälä, 'Depressive symptoms in adolescence as predictors of early adulthood depressive disorders and maladjustment', *Evidence-Based Mental Health*, volume 6, issue 2, p. 60, May 2003.

9 Royal College of Psychiatrists, 'Depression', Public Education Editorial Board, 2010.

10 The Royal College of Psychiatrists 'Alcohol and Depression', Public Education Editorial Board, 2008.

11 BBC, 'Self-harm by children on increase', BBC News, 2 May 2008.

12 The Royal College of Psychiatrists, 'Self-Harm, Suicide and Risk: Helping People who Self-Harm', June 2010.

13 'Harmful Drinking: Alcohol and Self-harm', NHS Quality Improvement, Scotland, 2007.

14 Sane, 'The kids aren't all right', as reported in John Crace, the *Guardian*, Saturday 18 September, 2010.

15 'Number of children prescribed drugs for ADHD and depression rises', *Nursing Times*, 30 October 2009.

16 American Psychiatric Association, *Diagnostic and Statistical Manual of Mental Disorders*, Fourth Edition (DSM-IV), Washington, DC, American Psychiatric Press, 1994.

17 J. E. Fleming and D. R. Offord, 'Epidemiology of childhood depressive disorders: a critical review', *Journal of the American Academy of Child and Adolescent Psychiatry*, volume 29, issue 4, pp. 571–80, 1990.

18 Royal College of Psychiatrists, 'Depression', Public Education Editorial Board, 2010.

19 Sara Eleoff, 'Divorce Effects on Children: An Exploration of the Ramifications of Divorce on Children and Adolescents', The Child Advocate, http://www.childadvocate.net/divorce_effects_on_children.htm), 2008.

20 Esme Fuller-Thompson, 'Is There a Link Between Parental Divorce During Childhood and Stroke in Adulthood? Findings from a Population Based Survey', annual meeting of the Gerontological Society of America, New Orleans, 22 November 2010.

21 P. R. Amato and J. Cheadle, 'The long reach of divorce: Divorce and child well-being across three generations', *Journal of Marriage and Family*, volume 67, issue 1, pp. 191–206, February 2005.

22 M. Fe Caces, et al., 'Alcohol Consumption and Divorce Rates in the United States'; *Journal of Studies on Alcohol*, volume 60, 1999; Mishcon de Reya/OnePoll, as reported by the *Daily Telegraph*, 'Divorces cause children to turn to alcohol', 17 June 2009.
23 S. S. Luthar and S. J. Latendresse, 'Children of the Affluent. Challenges to Well-Being', *Current Directions in Psychological Science*, volume 14, issue 1, pp. 49–53, 2005.
24 Tania Murray Li, 'Working Separately But Eating Together: Personhood, Property, and Power in Conjugal Relations', *American Ethnologist*, volume 25, issue 4, pp. 675–94, 1998.
25 The National Center on Addiction and Substance Abuse at Columbia University, www.casafamilyday.org, 2008.
26 The National Center on Addiction and Substance Abuse at Columbia University, 'The Importance of Family Dinners. National Center on Addiction and Substance Abuse at Columbia', statements reported by Fox News – Health: 'Family Meals Help Teens Avoid Smoking, Alcohol, Drugs', www.foxnews.com/story, 2005; the National Center on Addiction and Substance Abuse at Columbia University, www.casafamilyday.org, 2008.
27 National Family Month, 'National family month 2006 to provide opportunities to tackle top health, safety and social issues facing families today', http://www.nationalfamilymonth.net, 8 May 2006.
28 Seung-Schik Yoo, et al., 'The human emotional brain without sleep – a prefrontal amygdala disconnect', *Current Biology*, volume 17, issue 20, pp. R877–R878, 23 October 2007.
29 T. Roehrs and T. Roth, 'Sleep, sleepiness, and alcohol use', *Alcohol Research & Health*, volume 25, issue 2, pp. 101–9, 2001.
30 National Institute of Drug Abuse, 'Can Physical Activity and Exercise Prevent Substance Use: Promoting a Full Range of Science to Inform Prevention', 5–6 June 2008, Bethesda, MD.
31 F. B. Ortega, et al., 'Physical fitness in childhood and adolescence: a powerful marker of health', *International Journal of Obesity*, volume 32, pp. 1–11, 2008.
32 Chris J. Riddoch, et al., 'Objective measurement of levels and patterns of physical activity', *Archives of Disease in Childhood*, volume 92, issue 11, pp. 963–9, November 2007; Basterfield, et al., 'Surveillance of physical activity in the UK is flawed: validation of the Health Survey for England physical activity questionnaire', *Archives of Disease in Childhood*, 2008 (published online first, 9 September 2008; DOI:10.1136/adc.2007.135905).
33 Gavin Sandercock, et al., 'Twenty-metre shuttle run test performance of English children aged 11–15 years in 2007: Comparisons with international standards', *Journal of Sports Sciences*, volume 26, issue 9,

pp. 953–7, July 2008; Gavin Sandercock, et al., 'Ten year secular declines in the cardiorespiratory fitness of affluent English children are largely independent of changes in body mass index', *Archives of Disease in Childhood*, volume 95, pp. 46–7, 2010.

34 X. Sui, et al., 'Prospective study of cardiorespiratory fitness and depressive symptoms in women and men', *Journal of Psychiatric Research*, volume 43, issue 5, pp. 546–52, February 2009.

## Chapter 8

1 Coroners and Procurators Fiscal, 'Road Casualties Great Britain 2007', Department for Transport, published September 2008.

2 C. M. Farmer and A. F. Williams, 'Temporal factors in motor vehicle crash deaths', *Injury Prevention*, volume 11, pp. 18–23, 2005.

3 David F. Preusser, et al., 'Pedestrian crashes in Washington, DC and Baltimore', *Accident Analysis & Prevention*, volume 34, issue 5, pp. 703–10, September 2002.

4 D. Levitt and Stephen J. Dubner, *SuperFreakonomics: Global Cooling, Patriotic Prostitutes, and Why Suicide Bombers Should Buy Life Insurance*, Allen Lane, 2009.

5 P. Cummings P, et al., 'Changes in traffic crash mortality rates attributed to use of alcohol, or lack of a seat belt, air bag, motorcycle helmet, or bicycle helmet, United States, 1982–2001', *Injury Prevention*, volume 12, issue 3, pp. 148–54, June 2006.

6 'Road Casualties Great Britain 2007', Department for Transport, published September 2008.

7 G. Li, et al., 'Use of of alcohol as a risk factor for bicycling injury', *JAMA*, volume 285, pp. 893–6, 2001.

8 Insurance Institute for Highway Safety, 'Fatality Facts: Bicycles based on analysis of data from the US Department of Transportation's Fatality Analysis Reporting System', 2006.

9 L. Nicaj, et al., 'Bicyclist Fatalities in New York City: 1996–2005', *Traffic Injury Prevention*, volume 10, issue 2, pp. 157–161, 2009.

10 T. Varga, et al., 'Influence of alcohol: 114 hospitalised victims of traffic accidents', *Forensic Science International*, volume 103, supplement 1, pp. S25–S29, 16 August 1999.

11 Hannah Summers, 'Ben Kinsella killers jailed for life at Old Bailey', the *London Paper*, 12 June 2009; Helen Pidd, 'Ben Kinsella killers sentenced to life', *Guardian*, 12 June 2009; England and Wales Court of Appeal (Criminal Division) Decision, Case AM & Ors, R v [2009] EWCA Crim 2544, 13 November 2009.

12 Andrew Karmen, *Crime Victims: An Introduction to Victimology* (p.103), Cengage Learning, 2009.

13 The Alcohol Research Consortium, www.gla.ac.uk/schools/dental/researchactivities/biotechnologyandcraniofacial/researchactivities/alcohol-relatedfacialinjuries; C. A. Goodall, et al., 'Nurse-delivered Brief Interventions for Hazardous Drinkers with Alcohol-related Facial Trauma: A Prospective Randomised Controlled Trial', *British Journal of Oral Maxillofacial Surgery*, volume 46, issue 2, pp. 89–176, 2008.

14 Hannah Fletcher, 'Chef Peter Bacon cleared of raping solicitor so drunk she could not remember sex', *The Times*, March 27, 2009.

15 Stern Review, Equalities Office, Home Office, Crown Copyright 2010.

16 Legal Alliance, http://www.thelegalalliance.co.uk/Legal-Alliance-school-sued-after-drunken-student-falls-from-window-386.aspx, June 2009.

17 'Fall victim was drinking, police say', *North West Florida Daily News*, 3 April 2010; 'Police: Notre Dame recruit Matt James "drunk and belligerent" on fatal fall from motel balcony', *NY Daily News*, 3 April 2010.

18 'Drunk teenager died falling from hotel balcony at works party, hears inquest', *This is Cornwall*, 17 July 2010.

19 'Drinkaware warns parents to avoid giving their children alcohol over the summer', press release, 30 June 2010.

20 'Teens feel ill-equipped to handle consequences of underage drinking', Red Cross, 13 September 2010.

21 Joanna Sugden, 'Parents concerned about students' gap-year safety', *The Times*, 8 June 2009.

22 Alcohol Concern, 'Right time, right place: Alcohol-harm reduction strategies with children and young people', 2010.

23 'Alcohol-related hospital admissions soar', *Nursing Times*, 1 September 2010; 'Alcohol-related hospital admissions soar', *Health Service Journal*, 1 September 2010.

24 Alcohol 'is to blame for most weekend casualty admissions', *Daily Mail*, p. 13, 8 May 2009.

25 B. J. Casey, 'The storm and stress of adolescence: Insights from human imaging and mouse genetics', *Developmental Psychobiology*, Special Issue: Special Issue on Psychobiological Models of Adolescent Risk, volume 52, issue 3, pp. 225–35, 2010.

26 Laurence Steinberg, 'A dual systems model of adolescent risk taking', *Developmental Psychobiology*, Special Issue: Special Issue on Psychobiological Models of Adolescent Risk, volume 52, issue 3, pp. 216–24, April 2010.

27 Julia A. Graber, et al., 'Putting Pubertal Timing in Developmental Context: Implications for Prevention', *Developmental Psychobiology*,

Special Issue: Special Issue on Psychobiological Models of Adolescent Risk, volume 52, issue 3, pp. 254–62, 2010.

28 'Young people & alcohol: Young people's alcohol consumption & its relationship to other outcomes & behaviour', Department for Children, Schools and Families, National Centre for Social Research, project number: P2958, June 2010.

## Chapter 9

1 Department of Health, 'Guidance on the Consumption of Alcohol By Children and Young People', Sir Liam Donaldson, Chief Medical Officer, 17 December 2009.

2 US Department of Health and Human Services, 'The Surgeon General's Call to Action To Prevent and Reduce Underage Drinking', US Department of Health and Human Services, Office of the Surgeon General, 2007.

3 Mark A. Bellis, et al., 'Predictors of risky alcohol consumption in schoolchildren and their implications for preventing alcohol-related harm', *Substance Abuse Treatment, Prevention, and Policy*, volume 2, 2007, DOI:10.1186/1747-597X-2-15.

4 Adapted from 'Make a Difference: Talk to your child about alcohol', US Department of Health and Human Services, National Institutes of Health, NIH Publication Number 06-4314, 2009.

5 Caitlin Abar, et al., 'Myth of the forbidden fruit: The impact of parental modeling and permissibility on college alcohol use and consequences', May 2009; in Caitlin Abar and R. Turrisi (Chairs), 'Putting the "European Drinking Model" to the Test in the US: Evidence and Implications for Prevention', symposium presented at the annual meeting of the Society for Prevention Research, Washington, DC; Caitlin Abar, et al., 'The Impact of Parental Modeling and Permissibility on Alcohol Use and Experienced Negative Drinking Consequences in College', *Addictive Behaviors*, volume 34, issue 6–7, pp. 542–7, 2009.

## Chapter 10

1 American Psychiatric Association, *Diagnostic and Statistical Manual of Mental Disorders*, Fourth Edition, Washington, DC: the Association, 1994.

2 'Alcohol use disorders: diagnosis, assessment and management of harmful drinking and alcohol dependence', NICE guideline draft for consultation, June 2010.

3 N. A. Kaczynski and C. S. Martin, 'Diagnostic Orphans: Adolescents with Clinical Alcohol Symptomatology Who Do Not Qualify for DSM–IV Abuse or Dependence Diagnoses', paper presented at the annual meeting of the Research Society on Alcoholism, Steamboat Springs, CO, June 1995.
4 NCADD, http://www.ncadd.org has developed this set of questions for adolescents to reflect on.
5 TWEAK Scale, as described in P. Keogh, et al., 'Determining the effectiveness of alcohol screening and brief intervention approach in a young people's sexual health service', NHS Greater Glasgow and Clyde, 2008.
6 NICE, op. cit.

## Chapter 11

1 Drinkaware, House of Commons Memorandum by the Drinkaware Trust (AL 56), 8 April 2010 © Parliamentary copyright 2010; http://www.parliament.the-stationery-office.co.uk/pa/cm200910/cmselect/cmhealth/151/151we17.htm)
2 House of Commons Health Committee, 'Alcohol'. HC 151–I. London: The Stationery Office Limited, 2010.
3 ibid.
4 Ben Baumberg and Peter Anderson, 'The European Strategy on Alcohol: A Landmark and a Lesson', *Alcohol & Alcoholism*, volume 42, issue 1, pp. 1–2, 2007.
5 Thomas F. Babor, 'Alcohol research and the alcoholic beverage industry: issues, concerns and conflicts of interest', *Addiction*: Special Issue: The Alcohol Industry and Alcohol Policy, volume 104, pp. 34–47, February 2009.
6 Felicity Lawrence, 'McDonald's and PepsiCo to help write UK health policy. Exclusive', *Guardian*, 12 November 2010.
7 Home Office, 'Alcohol, young people and the law', Directgov, 2010.
8 Jamie Welham, 'Flip-flops for drunk revellers in Camden Town', *Camden New Journal,* 7 October 2010; 'Flip-flops, lollipops and coffee', *London Evening Standard*, p. 28, 12 October 2010.
9 Mark A. Bellis, et al., 'Teenage drinking, alcohol availability and pricing: a cross-sectional study of risk and protective factors for alcohol-related harms in school children', *BMC Public Health*, volume 9, p. 380, 2009; DOI:10.1186/1471-2458-9-380.
10 D. J. Nutt, et al., 'Drug harms in the UK: a multicriteria decision analysis', *Lancet*, Early Online Publication, 1 November 2010; DOI:10.1016/S0140 6736(10)61462-6.

# Resources

## UK

### Alcohol Concern

The national agency on alcohol misuse for England and Wales. Explains types of treatment services and general information about alcohol.
Telephone: 020 7264 0510
Website: www.alcoholconcern.org.uk/concerned-about-alcohol/alcohol-services

### NHS Care

NHS Care provides alcohol-addiction support and an online search function for local treatment options and services.
Website: www.nhs.uk/servicedirectories/Pages/ServiceSearchAdditional.aspx?ServiceType=Alcohol

## Drinkline

The NHS national drink helpline can provide local service information, as well as help and advice about alcohol problems.
Telephone: 0800 917 8282

## Addaction

UK-wide treatment agency, helping individuals, families and communities to manage the effects of drug and alcohol misuse.
Telephone: 020 7251 5860
Website: www.addaction.org.uk/

## Talk to Frank

National drugs awareness site for young people and parents/carers. Website has online services directory.
Telephone: 0800 776600
Website: www.talktofrank.com

## Alcoholics Anonymous

Telephone: 0845 769 7555
Email: help@alcoholics-anonymous.org.uk
Website: www.alcoholics-anonymous.org.uk/
Also has a web page for young people:
http://www.alcoholics-anonymous.org.uk/newcomers/?PageID=70

### Al-Anon

Support and understanding to the families and friends of current and former problem drinkers.
Confidential helpline: 020 7403 0888
Website: www.al-anonuk.org.uk/

### Parentline Plus

Advice and information for parents of children with an alcohol problem.
Freephone helpline: 0808 800 2222
Website: www.parentlineplus.org.uk

## Advice on Referrals for Private Therapy and Clinics

### Addiction Advisor

For free expert advice on which addiction treatment centre is most appropriate for your need.
Telephone: 0845 805 8342
Website: www.addictionadvisor.co.uk/addiction-centres/index.php

## Dry Out Now

For free expert advice on the full range of UK private alcohol rehab centres and out-patient programmes.
Telephone: 0845 118 0059
Website: www.dryoutnow.com/

## Action on Addiction

Works across the addiction field in research, prevention, treatment, professional education and family support.
Telephone: 0845 126 4130
Website: www.actiononaddiction.org.uk/

## Rehab Guide

Alcohol Addiction Treatment Locators in the UK, Scotland and Ireland.
Telephone: 01294 834 413
Website: www.rehabguide.co.uk/

## Counselling Directory

Provides the UK with a counselling support network, enabling those in distress to find a free counsellor close to them and appropriate for their needs. The site also features a section on alcohol addiction.
Telephone: 0844 8030 240
Website: www.counselling-directory.org.uk/alcohol.html

## Australia

### Alcoholics Anonymous Australia

Telephone: (02) 9599 8866
Website: www.aa.org.au

### Australian Drug Information Network

For alcohol and drug services.
Website: www.adin.com.au/content.asp?Document_ID=71

### Kids Helpline

A twenty-four-hour telephone and online counselling service for children and young people in Australia.
Telephone: 1800 55 1800
Website: www.kidshelpline.com.au

### Reach Out

Run by the Inspire Foundation, this is a web-based service providing information and a forum to help young people improve their mental health and wellbeing.
Website: http://au.reachout.com

### Somazone

A programme run by the Australian Drug Foundation (ADF), providing access to free health information online for young people.
Website: www.somazone.com.au

## New Zealand

### The Alcohol Advisory Council of New Zealand (ALAC)

Information, support, advice for parents or children.
Alcohol and drug helpline: 0800 787 797
Website: www.alcohol.org.nz/TeenagerInfo.aspx?PostingID=926

### Alcoholics Anonymous

Telephone: 0800 229 6757
Website: www.aa.org.nz/

### Youthline helpline services

Advice, support and information for parents and children.
Telephone: 0800 37 66 33
Website: http://youthline.co.nz/find-help/help-for-parentscaregivers.html

*Addiction/Dependency Services Locator:*
Email service for parents: parenttalk@youthline.co.nz
Website: http://youthline.co.nz/services-directory.html?catid=4

## South Africa

### Al-Anon Family Groups

Al-Anon Family Groups provides help and support to the families of problem drinkers.
Telephone: 0861 25 26 66
Website: www.alanon.org.za

### Alcoholics Anonymous

Telephone: 0861 HELP AA (435 722)
Website: www.aasouthafrica.org.za

### The South African National Council on Alcoholism and Drug Dependence (SANCA)

SANCA is a non-governmental organisation that aims to prevent and treat alcohol and drug dependence through the provision of drug help centres and treatment services.
Telephone: +27 11 781 6410 or 08614SANCA
Website: www.sancanational.org.za

# Index

# Index

# Index